THE CONTEMPORARY DIESEL SPOTTER'S GUIDE

BY LOUIS A. MARRE AND JERRY A. PINKEPANK

Editor: George H. Drury   Copy Editor: Marcia Stern   Art Director: Lawrence Luser

On the cover: Electro-Motive GP60 demonstrator EMD 5. Photo by Greg Sommers.

KALMBACH K BOOKS

PREFACE

Both the degree of change and the rate of change in the railroad motive power field have escalated dramatically since *Diesel Spotter's Guide Update* was published in 1979. The sharp downsizing of the rosters of most major railroads was accompanied by seismic upheavals in the increasingly unstable new locomotive market. Perhaps the most dramatic development was the January 1988 announcement by General Motors that its La Grange, Illinois, facility would be abandoned as a site of locomotive construction and whatever remained of its new locomotive business would emanate from the London, Ontario, plant — which would fit nicely into the La Grange parking lot. On the other hand, General Electric has come to dominate the field both technologically and in new sales. That those sales are now being made principally to just seven major U. S. railroad systems is another indication of how radical the changes have been since the *Update*.

This volume is intended to replace and expand upon the *Update*. It covers the period from 1972, when EMD introduced its Dash 2 line, to December 31, 1988. Included are expanded coverage of contract rebuildings and rebuilders and new coverage of such recent phenomena as large-scale lease turnbacks and massive resales, the virtual disappearance of freight electrification, and illustration of the remarkable proliferation of nuances among models. What the contemporary motive power world lacks in breadth is compensated for by the number and subtlety of variations.

Acknowledgments: Thanks are due to numerous railroad officials for research assistance. We particularly thank Joe McMillan of Santa Fe; S. R. Tedder of Ashley, Drew & Northern; Gordon B. Mott of CSX Transportation; Donald C. Gezon of Grand Trunk Western; Eric Post of Kansas City Southern; Marvin Black of Norfolk Southern; R. R. McClanahan of St. Louis-Southwestern; and George Cockle of Union Pacific. Others who provided generous assistance include Ken Ardinger, Warren Calloway, Peter Clark, Mac Connery, Curt Fortenberry, Herb Harwood, Gordon Lloyd Jr., Brian Matsumoto, Dr. John B. McCall, Pierre Patenaude, and Larry Russell.

A mixed-media contribution award must go to Gregory J. Sommers, who furnished not only a rainbow of color from which to choose a cover but also a virtual dial-a-unit negative supply and much moral support. To everyone else who assisted in what became a three-year project, our sincere thanks.

Louis A. Marre

Dayton, Ohio
March 5, 1989

Jerry A. Pinkepank

St. Paul, Minnesota
July 15, 1988

INTRODUCTION

In the early years of the diesel era — to 1950, let's say — little information about diesels was available to rail enthusiasts beyond builder identification, horsepower, and wheel arrangement. Statistical information such as tractive effort and weight had to be gleaned from press releases, and road classes and model designations got mixed up together — some diesels didn't even have model designations.

Railfans who were interested in diesels were in the minority. Most enthusiasts kept their attention on steam, at first because diesels would never replace steam locomotives, then because of the need to see, photograph, and ride behind steam locomotives before they were replaced by diesels.

The publication of Jerry A. Pinkepank's *Diesel Spotter's Guide* in 1967 by Kalmbach brought organization to an interest in diesels. The *Diesel Spotter's Guide* systematized the matter of diesel identification: If an Electro-Motive unit had three-axle Flexicoil trucks, eleven handrail posts, and two large radiator fans but lacked a turbocharger stack, it was an SD38. More than that, the book included pictures of all the diesel models since the beginning, and it brought order out of the chaos of model designations and specification numbers.

The *Diesel Spotter's Guide* was superseded by *The Second Diesel Spotter's Guide* in 1973; that was supplemented in 1979 by the *Diesel Spotter's Guide Update*, by Jerry A. Pinkepank and Louis A. Marre. This book begins its coverage at 1972, picking up where *The Second Diesel Spotter's Guide* leaves off.

The first half of this book deals with locomotives in as-built condition. They are arranged by builder: first Electro-Motive Division (of General Motors), for years the No. 1 builder; then General Electric, which replaced EMD in the top spot in 1983. Montreal Locomotive Works and its successor Bombardier Inc. follow EMD and GE. A few low-volume builders conclude the section.

The second half of the book covers locomotives that have been rebuilt, re-engined, de-engined — modified in some significant way from the way they were when they were delivered. The section titled "Jobs Locomotives Do" endeavors to explain how railroads select and assign locomotives. A glossary and an index finish the book.

Locomotives are classified as switcher, road-switcher, and cowl units (cab at one end; cab near one end; and full-width, non-structural body) and further as light, medium, and high-horsepower (less than 2000 h.p.; 2000-3000 h.p.; and greater than 3000 h.p.). A single date and a hyphen (such as 6/85-) in the column headed "Period produced" indicates a model still in production or still in the builder's catalog. A production figure of "None" indicates either "None yet" or "None ever," depending on the dates the model was produced.

Diesel spotting

Identifying a locomotive model requires first determining the builder. Most of the locomotives in North America today are products of either Electro-Motive or General Electric. The general appearance of the locomotives built by EMD and GE differ — the shape of the cab and the hood are the best clue.

Knowing builder and wheel arrangement — B-B (4 axles) or C-C (6 axles) — you should be able to turn to the appropriate pages of this book and home in on a specific model by using such characteristics as fans, louvers, and doors.

In most cases this book does not include roster information — lists of locomotives by the owning road's numbers. Charles W. McDonald's *Diesel Locomotive Rosters: U. S., Canada, Mexico* (Kalmbach, 1986) should serve most needs in that area.

The hazards of diesel spotting: A large railroad shop may be given some task to maintain a level of employment during a slack period. National Railways of Mexico's shop at San Luis Potosí is the undisputed champion when it comes to inventing a project to employ idle hands. Consider this example. Front to back it appears to be a GE low nose from a U28B, an EMD GP9-era cab, F7 dynamic brake fan and Far-Air filter grilles plus a piano crate to put them on, and an Alco RS-11 engine hood pretty much intact, perhaps with the original 251 engine inside, though that is a hazardous guess. Alco trucks support a frame which might have been built especially for the project, though faint suggestions of Alco heritage linger — excluding the vertical step wells. If necessity be the mother of invention, what can be said for a unit constructed by no impulse stronger than avoiding unemployment? Photo by Jim Herold.

ELECTRO-MOTIVE DIVISION, GENERAL MOTORS CORPORATION

Until recently, General Motors had the largest share of diesel-electric locomotive sales in the United States and Canada. Over 60 percent of the locomotives now in service in these two countries were manufactured by GM.

Plants: Plant No. 1, at La Grange, Illinois (actually in the city of McCook), was opened in 1934 and has been considerably expanded since. Plant No. 2, on the south side of Chicago, supplies parts and does subassembly work. From 1948 to 1954 switchers and some road-switchers were produced at Plant No. 3 in Cleveland, since sold to another GM division.

In 1950 a Canadian subsidiary, General Motors Diesel, Ltd., opened an assembly plant at London, Ontario. Today it is known as General Motors Diesel Division (GMDD). In January 1988 GM announced that the La Grange plant would be abandoned as a site of locomotive construction and its locomotive assembly business would be concentrated at London. Prime movers and some other components are currently built at La Grange, while frame and hood construction and final assembly are carried out in the smaller London plant.

History: In 1930 GM purchased Electro-Motive Corporation, a gas-electric railcar design and sales outfit, and the Winton Engine Co., Electro-Motive's chief supplier of gasoline and diesel engines. GM began to market standard diesel switchers and passenger engines through these subsidiaries in 1935. In 1939 the first mass-production road freight diesel, EMC model FT, was introduced and continued as the only U. S. freight diesel in production until 1945. EMC and Winton were merged by General Motors January 1, 1941, and became the Electro-Motive Division of General Motors Corporation.

GM maintained the leading position in U. S. diesel locomotive sales from 1945 until 1983, in some years selling as much as 89 percent of the diesel units purchased by U. S. railroads.

Engines: GM has always produced its own engines. The first locomotive diesel was the Winton model 201-A, cylinder bore and stroke $8'' \times 10''$. It was succeeded by the $8\frac{1}{2}'' \times 10''$ model 567 in late 1938. The 567 was replaced by the $9\frac{1}{16}'' \times 10''$ model 645 in 1966. The 710 engine, which has the same bore as the 645 but an $11''$ stroke, was introduced in 1984. All are 2-cycle, 45-degree V engines. The model numbers 567, 645, and 710 signify the displacement of one cylinder in cubic inches.

EMD SWITCHERS: 645 ENGINE, 1000 AND 1500 H.P., B-B

Model	AC/DC	H.P.	Cyls.	Length	Truck centers	Period produced	Approximate number of units sold U. S. A.	Canada	Mexico
SW1000	DC	1000	8	44′8″	22′0″	6/66-10/72	118	None	None
SW1001	DC	1000	8	44′8″	22′0″	9/68-6/86	151[1]	4[1]	19[1]
SW1500	DC	1500	12	44′8″	22′0″	7/66-1/74	807	None	None
SW1504	DC	1500	12	46′8″	23′0″	5/73-8/73	None	None	60
MP15DC	DC	1500	12	47′8″[2]	24′2″	2/74-11/80	237[1]	4[1]	5[1]
MP15AC	AC	1500	12	49′2″	24′8″	8/75-8/84	226[1]	4[1]	25[1]
MP15T	AC	1500	8	50′2″	24′8″	10/84-11/87	42	None	None

[1]Production totals are through December 31, 1987. Due to termination of 645 engine models, it is assumed no more will be built.
[2]Extended to 48′8″ in 1975 to allow larger stepwells. No change in truck centers.

When the 645E engine replaced the 567C and D engines in GM's catalog, the SW1000 and SW1500 emerged as evolutionary successors to the 8- and 12-cylinder 567C-engine switchers, the SW900 and SW1200. The main external differences from the 567-engine switchers are a higher cab profile, a larger-radius curve to the cab roof, and hood detail changes that include a headlight and number indicator

SW1000

The SW1000 is distinguished from the SW1500 by having only one stack instead of two, and by the short hood-top radiator opening which does not extend over the sandbox filler as it does on the SW1500. Photo by Vic Reyna.

fairing reminiscent of the SW1200RS in Canada, a grid-protected space around the ends of the radiator cores at the front of the top of the hood (the grid wraps around the hood top), and a new style grid over the front radiator fan shutters at the front of the hood.

The new cab profile caused clearance problems for industrial customers, and therefore EMD introduced the SW1001 with a lower cab much like that of previous EMD switchers. Only 21 SW1001s were produced through the end of 1972, when EMD apparently decided that the small market for the 8-cylinder switcher did not warrant two designs in the catalog and dropped the SW1000. Interestingly, there seems to have been no demand for a low-clearance 12-cylinder switcher even though several industrial users bought 12-cylinder switchers. There appears to be nothing that prevented EMD from using a low cab on these SW1500 and MP15-series locomotives, but none were ordered.

SW1500

Two stacks and a radiator opening at the top front of the hood which extends over the sandbox filler distinguish SW1500 from the SW1000. The Flexicoil trucks shown on Kentucky & Indiana Terminal 80 were optional; AAR type A

The SW1500 was the preferred switcher among railroad customers, but it faced a market limitation. More and more railroads wanted a locomotive that could perform in road service as well as in switching. The SW1500's short frame prevented the use of the Blomberg road truck and held down the weight of the locomotive. After in effect building a prototype in the form of 60 SW1504s for National Railways of Mexico in 1973, EMD solved the dual-service problem by introducing the lengthened MP15 in 1974. This allowed the B Flexicoil truck to be dropped from the catalog, except in the unlikely event of a customer ordering an SW1001 so equipped. The MP15 designation was changed to MP15DC when the MP15AC was introduced in 1975.

The MP15AC is, as its model designation implies, an MP15 fitted with an alternator and rectifier instead of the usual DC transmission, allowing a control system more compatible with larger EMD

SW1500

trucks were also used. K&IT 78, which has undergone a maintenance-related exchange of trucks, demonstrates that the two types of truck are interchangeable. Both photos by Louis A. Marre.

road locomotives. This model also has a road-engine-style radiator arrangement derived from the SD45T-2 and SD40T-2 models. Obviously aimed at the market niche occupied by aging GP7s and GP9s relegated to yards and local freights, the MP15AC's market penetration was limited by the lack of a toilet, which some state laws require on units used singly or as a lead unit in multiple-unit sets outside of yards. The GP15-1 of 1976 had a toilet and appropriated much of what would have been the MP15AC market.

The MP15AC continued to be ordered, though, in part because the GP15-1 was considerably more expensive unless the right type of trade-in was provided (GP7, GP9, F7, F9, or other model with Blomberg trucks). Thus, MP15ACs tended to be purchased by roads having non-EMD trade-ins, or no trade-in. The cheaper MP15DC continued to be ordered by roads that did not need the AC transmission or that needed locomotives to serve primarily as switchers.

A desire for fuel economy sparked the MP15T of 1984, which has a turbocharged 8-cylinder engine instead of the normally aspirated 12-cylinder engine of the MP15 models.

SW1500

Fordyce & Princeton No. 1503 is an SW1500 built in 1970. It was originally equipped with AAR switcher trucks and was fitted with Flexicoil trucks in 1982. Photo by Louis A. Marre.

The hood of the SW1001 is identical to that of the SW1000, but the distinguishing feature of the SW1001 is the low-profile cab. Photo by Ken Douglas.

SW1001

Virtually identical in appearance to the foot-longer MP15DC, the SW1504 is found only on the National Railways of Mexico, which has (at this writing) no MP15DCs. However, the prominent louvered air filter box behind the rear stack is unique to the SW1504. The use of the Blomberg B truck instead of the Flexicoil B or AAR type A truck provides another visual distinction from the SW1500. Photo by Jim Herold.

SW1504

MP15DC

The MP15DC is distinguished from the otherwise-similar SW1500 by having Blomberg B road trucks instead of Flexicoil B or AAR type A trucks. The air-filter box behind the rear stack (not found on some MP15DCs) also distinguishes the model from the SW1500, and the lack of louvers in this box distinguishes the MP15DC from the SW1504. Missouri Pacific 1549 has the modified Blomberg truck with rubber instead of elliptical springs, and shock absorbers in the R1, R3, L2, and L4 positions. Photo by James B. Holder.

MP15DC

Southern 2348 has an experimental muffler (which became standard after January 1, 1980, to satisfy EPA requirements) obscuring the dual stacks, and a bevel at the rear of the filter box. It also has conventional Blomberg B trucks instead of the modified version used on MP 1549. This unit also has the enlarged stepwells used after March 1975. Note that the coupler pocket has been almost "swallowed" by the end sill, compared to MP 1549, built in February 1975. Photo by Warren Calloway.

Missouri Pacific 1362, from an order of 20 MP15DCs delivered in January 1982, exhibits the production version of the stack muffler, with the stacks spaced normally instead of being close together as on the Southern prototypes. Extended coupler pockets indicate these units may have the old stepwell width, perhaps to accommodate the plow pilot, as they measure 48′8″ over pulling faces. Photo by Greg Sommers.

MP15DC

The MP15AC is distinguished from the MP15DC by the side-intake radiator air flow arrangement, illustrated by CSX 1156. Photo by Greg Sommers. **MP15AC**

MP15AC

Long Island's MP15ACs, such as 169 (from an order of 23 built in March and April 1977), are used regularly in passenger service, although they were purchased primarily for way freights. The presence of a headend-power/control-cab unit on the other end of the train, conventional LIRR practice, allows the MP15ACs to perform this duty without special passenger equipment. Photo by Charles Trapani.

MP15AC

MKT 56, from an order of four delivered in October 1980, exhibits the stack muffler that became mandatory that year. Photo by Louis A. Marre.

Seaboard System 1200, from the inaugural order of 14 delivered in October and November 1984, is an MP15T. All MP15Ts were built for Seaboard, except the last one, built in November 1987 for Dow Chemical. The most certain distinguishing feature from the MP15AC is the turbocharger stack. However, the number indicator enclosure is bigger (compare it with the MP15AC coupled behind) and on the right side of the unit is a "free flow" blower duct. Photo by Tom King.

MP15T

EMD LIGHT ROAD-SWITCHERS: 645E ENGINE, 1500 H.P., B-B

Model	AC/DC	H.P.	Cyls.	Length	Truck centers	Period produced	Approximate number of units sold U. S. A.	Canada	Mexico
GP15-1	DC	1500	12	54'11"	29'9"	6/76-3/82	310	None	None
GP15AC	AC	1500	12	54'11"	29'9"	11/82-12/82	30	None	None
GP15T	AC	1500	8	54'11"	29'9"	10/82-4/83	28	None	None

Production totals are through December 31, 1987. Due to cessation of 645 engine production, it is assumed no more will be built.

The GP15-1 evolved from the MP15AC as an attempt by EMD to solicit orders based on GP7 and GP9 trade-ins. The locomotive was designed to use components from these trade-ins, and when such trade-ins were furnished, the price was competitive with contractor or railroad-shop rebuild jobs. The increase in length over the MP15AC provided better weight distribution and better tracking at road speeds, and allowed room for a toilet and a road-engine-style cab with short hood for collision protection.

An aberration in GP15 evolution was the GP15AC, built only for Missouri Pacific at the end of 1982. The usual objective of maximum reuse of components was compromised by using a new AR10 alternator on these locomotives instead of the trade-in's DC generator. At the same time, EMD began producing a fuel-efficient variant, the 8-cylinder, turbocharged GP15T, which also used the AR10. The GP15 series didn't sell well, because it was generally intended for services that don't warrant the purchase of new locomotives — trains for which old-soldier GP7s and GP9s are quite adequate.

GP15-1

Chicago & North Western and St. Louis-San Francisco ordered GP15-1s without inertial air filters, hence the louvers at the top of the hood behind the cab for conventional air filters. Photo by Gordon B. Mott; detail photo by Louis A. Marre.

Missouri Pacific's version with inertial air filters is represented by MP 1677. Photo by Louis A. Marre.   **GP15-1**

GP15AC

MP's GP15ACs now wear yellow paint and Union Pacific lettering, like 1737, and are equipped with MP's own "liberated exhaust" four-stack manifold. Photo by Greg Davies.

GP15AC

Missouri Pacific 1715-1744 are the only GP15ACs built. They were delivered in MP's blue livery, as illustrated by 1717. Photo by John B. McCall.

GP15T

Chesapeake & Ohio 1500-1524, the first GP15Ts, were the first GP15s with dynamic braking. The turbocharger stack is hidden by the muffler mandated by the Environmental Protection Agency, but the larger, double-unit radiators and the enlarged radiator air intake opening identify the unit. The units have the same numbers on the CSX roster. Photo by Louis A. Marre.

GP15T

The only other GP15Ts are Apalachicola Northern 720-722, built in April 1983. They lack dynamic brakes. Photo by Greg Sommers.

EMD MEDIUM ROAD-SWITCHERS: 645E ENGINE, 2000 and 2300 H.P., B-B

Model	AC/DC	H.P.	Cyls.	Length	Truck centers	Period produced	Approximate number of units sold U. S. A.	Canada	Mexico
GP38	DC	2000	16	59'2"	34'0"	1/66-12/71	466	21	6
GP38AC	AC	2000	16	59'2"	34'0"	1971	240	None	None
GP38-2	AC	2000	16	59'2"	34'0"	1/72-	1801[1]	254[1]	133[1]
GP38P-2[2]	AC	2000	16	59'2"	34'0"	6/75	None	None	20[1]
GP39	AC	2300	12	59'2"	34'0"	5/69-7/70	21	None	None
GP39DC	DC	2300	12	59'2"	34'0"	6/70	2	None	None
GP39-2	AC	2300	12	59'2"	34'0"	8/74-	249[1]	None	None

[1]Production totals are through December 31, 1987. With cessation of 645 engine production, it is assumed no more will be built.
[2]GP38P-2 designation is not official; these units were ordinary GP38-2s with a high short hood housing a steam generator.

When the 645E engine replaced the 567C and 567D engines in GM's catalog, the GP38 was introduced as a 3-foot-longer version of the otherwise-similar 567-powered GP28. Although the GP28 had registered only limited sales during the 20 months in which it was produced, EMD perceived that there was a market for a less-than-3000 h.p. locomotive, especially one without a turbocharger (which is a maintenance headache if not needed for higher output). The GP38 was the result.

However, EMD also felt that some customers would rather take the turbocharger and four fewer power assemblies, so it offered the GP39 as an alternative. As it turned out, only Chesapeake & Ohio (20 units: Nos. 3900-3919), Atlanta & St. Andrews Bay (1 unit: No. 507), and Kennecott Copper (2 units: Nos. 1 and 2) opted for the turbocharger. Kennecott ordered its units fitted with a DC generator instead of an alternator. When the Dash 2 line appeared in January 1972, no mention was made of a GP39-2.

Subsequently, however, customers requested the model, which EMD had indicated would be available on special order. The first customer was Santa Fe, which wanted the turbocharger for operation at high altitudes — between Pueblo and Denver, Colorado, for example. Non-turbocharged locomotives lose considerable output at high altitudes because of the thin air, and may smoke objectionably.

The next customer was the Reading, to which the old argument of four fewer power assemblies must have appealed — Reading had no altitude problems. Fuel economy soon became the leading argument for the GP39-2, and it went on to achieve a respectable production figure.

The GP38-2 was ultimately a runaway favorite, and GP38s and GP38-2s soon acquired exceptional reputations for high availability and low operating cost. National Railways of Mexico ordered 20 GP38-2s equipped with steam generators for passenger service, and they carried an unofficial model number of "GP38P-2." Since there was no change in the locomotive except for the steam generator, there was no basis for a separate model number — except that 1975 was such a late date to be building a steam-equipped locomotive that the distinction seemed worthwhile. High-short-hood versions were also furnished to Norfolk & Western and Southern to satisfy their penchants for a locomotive that crews would not object to operating long-hood forward.

It was reported in 1986 that GMDD had proposed to CN a "GP69," essentially a GP59 in a full-width body. In addition, GO Transit ordered 16 F59PHs, delivery of which began in August 1988. They are, in effect, GP59s in full-width bodies with head-end power for passenger-car heating and lighting.

One last model to be mentioned is the GP38AC, which was offered as an option to the regular DC model during 1971. There is no external difference between the AC and DC models. U. S. production of the GP38-2s ended with Texas-Mexican 867 in May 1985, but Canadian Pacific continued to order them from GMDD, receiving 50 units built in 1987.

Both GP38

Detroit, Toledo & Ironton 216 illustrates a GP38 with paper air filters but without dynamic braking. The two stacks (topped by spark arresters) instead of a single crosswise turbocharger stack distinguish the non-turbocharged GP38 from otherwise-similar models. The 567-engined GP28 is similar in appearance but 3 feet shorter; on a GP28 there is no space between the front end of the radiator and the first power assembly door from the rear (on which the letter ''I'' is painted). Since only 26 GP28s were built, 10 of them for Mexico, confusion does not often arise. Baltimore & Ohio 3845 is a GP38 with normal-range dynamic braking (extended-range braking would be indicated by an access door for contactors in the blister). B&O 3845 lacks paper air filters, which would appear as a box raised a little above the hood line (see the location of the box on DT&I 216). Here the twin stacks are barely visible in the form of triangular spark screens. The two fans over the radiators, however, remain visible, as distinct from three such fans on a GP40, or two such fans sandwiching a smaller fan on the 567-engined GP35. Both photos by Louis A. Marre.

Both GP38-2

No single spotting feature is totally reliable for distinguishing a GP38-2 from a GP38. Kansas City Southern 4001 is a GP38-2 with new-style Blomberg B trucks, paper air filter, and dynamic braking. Only the lack of a hinged battery box cover with latches and the stubbier radiator area with closely spaced fans reveal the locomotive as a Dash 2. However, the first few GP38-2s had the old radiator arrangement, and Burlington Northern GP38-2s were ordered with hinged battery box covers. Toldeo, Peoria & Western 2001 is a later GP38-2 with the paper air filter box faired down. This was first done specially on Long Island units delivered starting in January 1976. EMD then made it a standard item for GP38-2s equipped with paper air filters. The change in the radiator area is evident in comparing TP&W 2001 with DT&I 216. The stacks of TP&W 2001 are not visible but can be inferred, since a turbocharger stack would be visible from this angle. The oblong water-level sight glass below and to the right of the radiator is a right-side-only detail on most Dash 2s. Upper photo by James B. Holder; lower photo by Louis A. Marre.

22

Both GP38-2

Seaboard System 4141 is a GP38-2 with the Blomberg High Traction truck (note the rubber spring and shock absorbers), paper air filters, and extended-range dynamic braking. Canadian National 5592 is equipped with a so-called Comfort Cab. CN's 51 GP38-2s so equipped are the only Comfort Cab GP38-2s. Delivery of the units began in 1973. Upper photo by Greg Sommers; lower photo by Larry Russell.

GP38P-2

The 20 steam generator-equipped GP38-2s built for NdeM in 1975 were unofficially tagged GP38P-2s, but except for the steam generator they are the same as high-hood GP38-2s built for N&W and Southern. Photo by Tom Chenoweth; Larry Russell collection.

GP39

Atlanta & St. Andrews Bay 507 illustrates the position of the single, fat turbocharger stack set crosswise on the roofline where the paper air-filter box is located on GP38s so equipped. The combination of this stack with just two fans indicates a GP39 or GP39-2. The distinctions for the Dash 2s are similar to those for the GP38, but because C&O 3900-3919 (now CSX 4280-4299), A&StAB 507, and Kennecott 1 and 2 are the only GP39s, the opportunity for confusion does not often arise. Only the C&O-CSX units have dynamic brakes. Photo by Greg Sommers.

Delaware & Hudson 7613 exhibits the turbocharger stack and is equipped with high-traction trucks. Interestingly, early GP39-2s such as D&H 7613 used the old GP38 radiator (spanning all or part of six door panels), while units built in 1977 or later had the GP38-2 version spanning just four door panels. Apparently the smaller radiator was at first thought to be insufficient for the cooling demands of 300 more horsepower, but experience must have proved otherwise. At the same time as the radiator change was made, the prime mover was relocated about 3 feet to the rear and hood openings were reoriented accordingly. Photo by J. R. Quinn.

Both GP39-2

Kennecott Copper Company took delivery of 11 GP39-2s in January 1977 with a special cab configuration and high underclearance for rocks found on the track in its Bingham (Utah) open pit mine. Subsequently, 10 more were acquired in 1978, and a final 7 in 1980. The decline in copper mining caused Kennecott to sell 9 units to Missouri-Kansas-Texas; MKT rebuilt the cabs to conventional configuration. These Kennecott GP39-2s were the first with the compact radiator and the engine shifted 3 feet to the rear, as mentioned above. Photo by Paul C. Hunnell.

Ex-Kennecott Copper GP39-2 as it looks on MKT. A contractor rebuilt the cab, but the extra clearance under the fuel tank is still evident. Photo by Louis A Marre. **GP39-2**

ATSF 3696 shows the three-foot greater distance between the cab and the traction motor blower duct common to all post-1977 GP39-2 production. Compare it to D&H 7613 on page 25. Photo by Greg Sommers.

GP39-2

MKT 364 is one of 20 units (360-379) delivered between March and June 1984. They were the last GP39-2s built, and there had been a gap of almost three years in GP39-2 production. The unit has the same external identification features as the GP49. Photo by George Cockle.

GP39-2

There have been 121 GP38-2s built since the free-flow blower duct was adopted in December 1982, 115 of them for Canadian Pacific (3021-3040, built March-June 1983; 3041-3085, September 1985 through February 1986, exemplified by 3078; and 50 more, 3086-3135, ending in 1987). Photo by Curt Fortenberry Jr.

GP38-2

The other late GP38-2s are Soo 4451 and 4452, built in March 1983; Ontario Northland 1808 and 1809, September 1984; Devco 228, July 1983; and Texas-Mexican 867, May 1985 (the last domestic GP38-2). Photo by Gordon Lloyd Jr.

GP38-2

EMD MEDIUM ROAD-SWITCHERS: 645 ENGINE, 2600 and 2800 H.P., B-B

Model	AC/DC	H.P.	Cyls.	Length	Truck centers	Period produced	Approximate number of units sold U. S. A.	Canada	Mexico
GP39X	AC	2600	12	59'2"	34'0"	11/80	6	None	None
GP49	AC	2800	12	59'2"	34'0"	8/83-5/85	9[1]	None	None

[1]Production totals are as of December 31, 1987. It is assumed no more will be built.

In November 1980, Southern Railway received GP39X locomotives 4600-4605, the prototypes of a higher-horsepower intermediate model — in effect a 12-cylinder edition of the GP50. After these six units had been tested sufficiently, the model went into the catalog as the GP49, rated at 2800 h.p. rather than the 2600 h.p. of the first units. Only the Alaska Railroad purchased the GP49, however, in two orders (August 1983 and May 1985). The railroads' lack of inter-est in this model is partially explained by the low level of locomotive orders in this period, but also by the fact that the tried-and-true GP39-2 remained in the catalog. Indeed, MKT bought twenty GP39-2s (360-379) after the GP49 was in the catalog, and they are externally indistinguishable from the GP49. Creation of the 710-engined GP59 left the Southern's GP39X units (now designated GP49s by EMD and the owner) and the Alaska units as the only examples.

GP39X

The prime identification feature of the GP39X (apart from the fact that only six exist, Southern 4600-4605) is the presence of large GP50-style radiators with only two fans (rather than three). Since 1982, these six units have been rated at 2800 hp and are GP49s for all practical purposes. Photo by Marvin Black.

Both GP49

Alaska 2801 is from the initial order of four units, 2801-2804, delivered in August 1983. Alaska 2805 is from the second order, 2805-2809, delivered in May 1985. The L-shaped windshield is an option EMD has had in the catalog since the SD45 prototypes of 1965. The hoods over the central air intake and the vertical snow baffles on top are custom features for the Alaska Railroad's harsh environment. The "free-flow" blower duct is a left-side-only feature. As on the GP39X, there are two radiator fans, but one of them on the Alaska units is concealed by a winterization hatch. Both photos by Curt Fortenberry Jr.

EMD MEDIUM ROAD-SWITCHER: 710 ENGINE, 3000 H.P., B-B

Model	AC/DC	H.P.	Cyls.	Length	Truck centers	Period produced	Approximate number of units sold		
							U. S. A.	Canada	Mexico
GP59	AC	3000	12	59'2"	34'0"	6/85-	3	None	None

[1]Production totals are as of December 31, 1988.

EMD plans to standardize future locomotive production on its 710 engine, and the GP59 will be the successor to the GP49. As of this writing, demonstrators EMD-8 through EMD-10 are the only units. Alaska's last order of GP49s came after the introduction of the GP59. The demonstrators were sold to Norfolk Southern in October 1986.

It was reported in 1986 that GMDD had proposed to CN a "GP69," essentially a GP59 in a full-width body. In addition, GO Transit ordered 16 F59PHs, delivery of which began in August 1988. They are, in effect, GP59s in full-width bodies with head-end power for passenger-car heating and lighting.

GP59

EMD-9 is one of three GP59 demonstrators. Photo by Marvin Black.

EMD RAIL-GRINDING TRAIN POWER UNIT: 645E ENGINE, 2000 H.P., B-B

Model	AC/DC	H.P.	Cyls.	Length	Truck centers	Period produced	Approximate number of units sold U. S. A.	Canada	Mexico
F40PH-2M	AC	2000	16	56'2"	33'0"	3/82-11/85	4		

On special order from Speno Rail Services, a rail resurfacing contractor headquartered in Syracuse, New York, EMD modified the carbody of the standard F40PH to accommodate the machinery of a GP38-2, a non-turbocharged 16-645E engine rated at 2000 h.p. Upon delivery, Speno further modified the carbody in its shops, moving the windshields out to the front of the nose and eliminating the "snout" entirely. The units are deployed on each end of two multicar rail-grinding trains and are numbered in sequence with their respective sets of equipment. Currently they are 101, 109, 201, and 212, though these numbers change as the trainsets are reconfigured.

Speno units 201 and 212 as they appeared and were numbered in late 1987. Aft of the cab door, they are recognizable as F40s. Both photos by Greg Sommers.

F40PH-2M

Between the two orders of F40PH-2M units, Speno rebuilt four ex-Conrail GP38s into similar units. As the units are renumbered to correspond to the trainsets which they are powering, identification by number can be confusing. MW unit 201 is shown here in September 1985. It and a mate were acquired in March 1984, followed by a third in December 1984. Spotting features include the full GP38 fuel tank, and two radiator sections rather than the three on the F40s. The lack of a "brow" on the front is conspicuous, but Speno has reconfigured the noses on all its power units at least twice. Photo by Gordon Lloyd Jr.

Though technically not a locomotive because it does not pull a trailing load, Speno's RSA-1 (Rail Surface Analyzer) is another interesting product of the company's Syracuse shop. Somewhat similar in function to the familiar Sperry rail analyzer cars, the unit is considerably larger: 69' long and approximately 65 tons. It is powered by one 450 h.p. Cummins diesel and rides on trucks from a Budd RDC. Photo by Larry Russell.

EMD HIGH HORSEPOWER ROAD-SWITCHERS: 645E ENGINE, 3000 H.P., B-B

Model	AC/DC	H.P.	Cyls.	Length	Truck centers	Period produced	Approximate number of units sold U.S.A.	Canada	Mexico
GP40	AC	3000	16	59'2"	34'0"	11/65-12/71	1201	24	18
GP40TC	AC	3000	16	65'8"	40'6"	11/66-12/66	None	8	None
GP40P	AC	3000	16	62'8"	37'3"	12/68	13	None	None
GP40-2	AC	3000	16	59'2"	34'0"	4/72-11/86	812[1]	275[1]	44[1]
GP40P-2	AC	3000	16	62'8"	37'3"	11/74	3	None	None

[1]Production totals are through December 31, 1987. With the discontinuance of 645 engine production, it is assumed no more will be built.

As horsepower per unit climbed above 3000 in the mid-1960s, North American railroads began to show a preference for 6-motor units. The first railroad to select C-C diesels for fast freight was the Atlantic Coast Line in 1963. Other railroads soon followed ACL's lead, largely because of the slipperiness of high-horsepower B-Bs but also because of a spate of traction-motor problems on the GP40. EMD at first steered railroads away from 6-axle units, pointing out that the performance curves were no different above 12 mph, but problems with the B-Bs spoke more eloquently than performance curves.

Sales of the GP40 were brisk initially with a continuation of ear-lier unit-reduction orders for GP20s, GP30s, and GP35s, but the momentum soon ran down. By the time the Dash 2 line was introduced, the downward trend was established. (As it turned out, the GP40-2 eliminated many of the causes of complaint about the GP40.) Had it not been for orders from Chessie System (162 of the 323 U.S. units through 1977) and Canadian National, which bought most of the Canadian output, the model would have been a poor seller. A revival of interest came with the fuel crisis of 1979, however, and 489 GP40-2s were built from 1978 to 1985.

GP40

The combination of B-B wheel arrangement and three large radiator fans spells GP40, whether or not the fat turbocharger stack is visible. Texas, Oklahoma & Eastern D-16 and D-14 are examples without dynamic braking. Photo by Louis A. Marre.

The GP40TC was designed and built for the Toronto commuter trains of Government of Ontario Transit. Number 9803 shows the current appearance of the GP40TCs after the rear hood was rebuilt to silence the 500-kw head-end lighting generator. GO Transit has since acquired Comfort Cab and conventional GP40-2s that are used with auxiliary power cab units converted from F units, avoiding the need for any Dash 2 version of the GP40TC. The GP40TCs were delivered as 600-607, then renumbered 9800-9807 and later 500-507. These units were replaced by F59PHs in 1988 and sold to Amtrak. Photo by Ken Douglas.

GP40TC

Both GP40P

Thirteen GP40Ps numbered 3671-3683 were built for Central Railroad of New Jersey commuter trains. They had a longer frame and a projection at the rear containing a steam generator. Even with the longer frame, the radiators had to be mounted SD45-style. These units are now being stripped of their steam generators in favor of head-end-power packages for electric train heating and lighting. New Jersey Transit calls the rebuilds "GP40PHs." Numbers are NJT 4100-4112. CNJ 3680 photo by Ken Douglas; NJDOT 4111 photo by Gordon B. Mott.

Both GP40-2

Boston & Maine 311 is a GP40-2 with the rubber-sprung Blomberg M truck but without dynamic brakes. The bolted battery box cover is the only real indication that is a Dash 2. Photo by J. R. Quinn. Canadian National 9416 is a Comfort Cab version, also lacking dynamic braking. GO Transit also has some of the Comfort Cab version. Photo by Pierre Patenaude.

GP40-2

Baltimore & Ohio 4143 has dynamic brakes, but lack of a small, square access door in the dynamic brake blister ahead of the resistor grids shows it is not extended-range dynamic braking. The oval sight glass above the "H" in "Chessie" is the most reliable spotting feature to distinguish the GP40-2 from the earlier GP40. Photo by Louis A. Marre.

GP40-2

GP40-2s delivered from 1983 to 1986 had the free-flow blower duct on the left side. These units — Rio Grande 3129 and 3130, Southern Pacific 7240-7247, Cotton Belt 7248-7273, and Florida East Coast 430-434 — should not be confused with GP50s, whose larger radiator grilles distinguish them at a glance. D&RGW 3130 has extended-range dynamic braking and corrugated radiator grilles. Photo by Louis A. Marre.

Florida East Coast 434, delivered in December 1986, was the last GP40-2 built. Photo by Greg Sommers.   **GP40-2**

Aside from Dash 2 changes, Southern Pacific GP40P-2s 3197-3199 differed from their CNJ counterparts mainly in having smaller fuel and water tanks, extended-range dynamic braking, and cab air conditioning. After being displaced from Bay Area commuter trains by Caltrans F40PHs, the GP40P-2s were stripped of their steam generators and renumbered 7600-7602. Photo by Vic Reyna.

GP40P-2

EMD HIGH HORSEPOWER ROAD-SWITCHERS: 645F ENGINE, 3500 H.P., B-B

Model	AC/DC	H.P.	Cyls.	Length	Truck centers	Period produced	Approximate number of units sold U.S.A.	Canada	Mexico
GP40X[1]	AC	3500	16	60'2"	35'0"	12/77-6/78	23	None	None

[1]Duplicates a model number used unofficially for GP40 prototype EMD No. 433A of 1965.

The GP40Xs carried EMD's first F-series engines in regular production. They were prototypes for the "50" line. They achieved improved adhesion through a single-axle wheelslip detection and control system. (GE's version, as used on five BN U30Cs, was called SAWS, for Single Axle Wheelslip System, and is now called the Sentry system by GE. The ancestor of both is the ASEA system used on Swedish electric locomotives.) Roster of GP40Xs: Santa Fe 3800-3809; Southern 7000-7002; Southern Pacific 7200, 7201, 7230, and 7231; and Union Pacific 90-95 (formerly 9000-9005).

GP40X

Union Pacific's GP40Xs ride on HT-B trucks, as illustrated by UP 92. The units were delivered as 9000-9005 and renumbered 90-95. Photo by George Cockle.

Santa Fe 3805 rides on the familiar Blomberg M truck. Santa Fe's GP40Xs are numbered 3800-3809. Photo by George R. Cockle; collection of Larry Russell. **GP40X**

Southern's GP40Xs, 7000-7002, were equipped with Blomberg M trucks and, following Southern's usual practice, were built with a high short hood and the long end designated as front. Here 7000 is coupled to GP50 No. 7032, affording a radiator area comparison. Photo by Marvin Black.

GP40X

GP40X

Southern Pacific 7200 and 7201 (RCE masters) and 7230 and 7231 (RCE remotes) had the HT-B truck, optional L-shaped windshield, and "elephant ears" designed to bring cooler track-level air up to the radiators when traversing tunnels. Photo by George R. Cockle.

EMD HIGH HORSEPOWER ROAD-SWITCHERS: 645F ENGINE, 3500 H.P., B-B

Model	AC/DC	H.P.	Cyls.	Length	Truck centers	Period produced	Approximate number of units sold U.S.A.	Canada	Mexico
GP50	AC	3500	16	59'2"	34'0"	5/80-11/85	278	None	None

The GP40-2 remained in production after the GP50 was introduced; indeed, more than 100 GP40-2s were built side by side with GP50s. Ironically no GP50s were built in 1986, but 2 GP40-2s were constructed for Florida East Coast. The reasons for the overlap in production appear to be, first, that the railroads wanted to fill out fleets with compatible power and, second, uneasiness about the horsepower increase from a maintenance viewpoint. In another era,

EMD might have told the railroads to "take it or leave it" with a new model, but with declining sales in the 1980s, there was a tendency to cater to the railroads' needs.

The GP50 has generally been purchased for service requiring high speed and reduced locomotive weight, such as TOFC/COFC trains. EMD considered both the GP40-2 and GP50 closed out as of the end of 1986 and replaced in its catalog by the GP60.

GP50

Chicago & North Western 5067 is from the initial GP50 production order, CNW 5050-5099. The bigger radiator grille is the main distinguishing feature of the GP50 compared to the GP40-2. Number 5067 exhibits the elongated blister used for extended-range dynamic brakes. Photo by Warren Calloway.

The second production order of GP50s was Southern 7003-7072, built between August and December 1980. **GP50**
Like the C&NW units, these locomotives have Blomberg M trucks, the old-style blower duct (this side only),
and a short blister indicating conventional-range dynamic braking. No GP50s were built with HT-B trucks. The
old-style blower duct appeared on all GP50s built through 1981. No GP50s were built in 1982, 1983, and 1984.
Photo by Louis A. Marre.

Missouri Pacific bought 10 GP50s at the end of 1980, Nos. 3500-3509, and another 20, 3510-3529, at the beginning of 1981. None had dynamic braking. Photo by Louis A. Marre.

GP50

GP50

GP50 deliveries resumed in 1985 after a 3-year lapse. Burlington Northern obtained 53 units, 3110-3162, and Santa Fe took 15, numbered 3840-3854. These units have the free-flow blower duct as shown on BN 3142 (left side only). Number 3142 has conventional-range dynamic braking, indicated by the short blister, and conventional leaf-spring trucks. Photo by Greg Sommers.

GP50

The last five units in the BN order, 3158-3162, had enlarged cabs for cabooseless operation, reducing nose length from the usual 88 inches to 65 inches. Number 3162 is shown coupled to a fuel tender. These enlarged-cab GP50s were also the last GP50s built, making No. 3162 the last of the model run. Both photos by Roger Bee.

EMD HIGH HORSEPOWER ROAD-SWITCHER: 710G ENGINE, 3800 H.P., B-B

Model	AC/DC	H.P.	Cyls.	Length	Truck centers	Period produced	Approximate number of units sold U.S.A.	Canada	Mexico
GP60	AC	3800	16	59'2"	34'0"	10/85-			

During 1985 and 1986, EMD fielded its 710G-engine prototypes, among them three GP60s, EMD 5, 6, and 7. These units featured rounded corners on the cab and the short hood to reduce wind drag, but the experiment was not carried into the first round of production units. In other respects the GP60 is identical in appearance to the GP50 and the GP59.

The only novelty in the first production units was their place of construction. As production began, EMD announced that it was moving most if not all of its assembly operations to the London, Ontario, plant of General Motors Diesel Division.

Redesigned builder's plates will henceforth appear on GM diesel locomotives, and "La Grange" may become just a name of a Chicago suburb as far as railroads are concerned.

GP60

The prototype units have conventional dynamic braking and rubber-spring Blomberg M trucks. Photo by Greg Sommers.

Production GP60s closely resemble GP50s and GP59s. The new EMD/GMD combined builder's plate, though not exactly a spotting feature, does distinguish the models. An X in the appropriate block tells where the unit — in this case Santa Fe 4004 — was built. Both photos by Greg Sommers.

GP60

EMD MEDIUM ROAD-SWITCHERS: 645E ENGINE, 2000 AND 2300 H.P., C-C

Model	AC/DC	H.P.	Cyls.	Length	Truck centers	Period produced	Approximate number of units sold U. S. A.	Canada	Mexico
SD38	DC	2000	16	65'9½"	40'0"	5/67-7/71	38	None	None
SD38AC	AC	2000	16	65'9½"	40'0"	6/71-10/71	14	1	None
SD38-2	AC	2000	16	68'10"	43'6"	11/72-6/79	74	7	None
SD39	AC	2300	12	65'9½"	40'0"	8/68-5/70	54	None	None
SDL39	AC	2300	12	55'2"	30'0"	3/69-11/72	10	None	None

Because of the nature of the six-motor, 2000-2300 h.p. locomotive as a special-application machine, sales have been scattered. Twenty months passed between the delivery of the last 567-engine version — only six SD28s were built, all in 1965, for Columbus & Greenville and Reserve Mining — and the first SD38. It was another 13 months from the last SD38 delivery to the first SD38-2, and from 1975 to 1979 only 16 units were built. The other 65 units were built between November 1972 and the end of 1975 — 33 in 1975.

The biggest customers for the units have been the U. S. Steel roads, which are characterized by slow, heavy trains. Of the 74 SD38-2s built in the U. S., 30 went to Bessemer & Lake Erie; Elgin, Joliet & Eastern; and Duluth, Missabe & Iron Range. Most of the re-

Bessemer & Lake Erie SD38AC No. 869, equipped with dynamic brakes and paper air filters, is indistinguishable from DC SD38s. The SD38 and SD38AC are distinguished from the SD28 by having 11 hand-rail stanchions along the long hood instead of 10. They are distinguishable from SD39s by the absence of the fat turbocharger stack, having instead two small stacks on the center line of the hood ahead of and behind the dynamic brake fans. Even if the stacks were not visible, the paper air-filter box, as on B&LE 869, is where a turbocharger stack would be, indicating that this is a nonturbocharged SD38 rather than a turbocharged SD39. Two large radiator fans distinguish SD38s, SD39s, and SD38-2s from similar SD40s and SD40-2s, which have three. The SD35 has two large fans with a small one between them. Photo by Gordon Lloyd Jr.

SD38AC

maining sales have been for hump locomotives for such roads as Southern Pacific; Louisville & Nashville; Chicago & North Western; Penn Central; Conrail; Detroit, Toledo & Ironton; and Frisco. During the last year of SD38 production the model was offered with AC transmission instead of DC as an option. All SD38s built thereafter were AC machines except for DT&I's last two.

EMD apparently intended that the SD39 replace the SD38 in the catalog, but as with the GP39 and the GP38, EMD found that many roads preferred units without a turbocharger. The SD39 sold as well as its catalog mates, the SD38 and the SD38AC (54 versus 53 units), but when the Dash 2 line was introduced January 1, 1972, no SD39-2 was listed in the catalog. EMD emphasized that it would continue to build units to customer order, but unlike the GP39-2, no customer demand for an SD39-2 developed.

Grand Trunk Western 6254 (ex-Detroit, Toledo & Ironton 254) is a DC-generator-equipped SD38 without dynamic brakes. Photo by Louis A. Marre.   **SD38 (DC)**

Like B&LE 869 and GTW 6254, Chicago & North Western 6657 has paper air filters. However, paper air filters are an option and therefore should not be depended on as a spotting feature for SD38s and SD38-2s. The oval water level sight glass just below the front of the radiator (right side only) and the bolted-on battery box covers are clues that this is a Dash 2. Photo by Greg Sommers.

SD38-2

Reserve Mining 1237-1245, built in October 1978, were among the last SD38-2s built. Only Yankeetown Dock 22, built in December 1978, and Frisco 296-299, built in June 1979, came later. The sloped air filter housing is the main feature distinguishing these units from earlier SD38-2s. Unlike Chicago & North Western 6657, the Reserve units have dynamic braking. Reserve Mining shut down its operations in 1986, and units 1237-1245 went on the secondhand locomotive market. Photo by Roger Bee.

SD38-2

Both SD39

Southern Pacific 5316 has extended-range dynamic brakes, extra lights, a large fuel tank, and a pilot plow. Photo by Vic Reyna. Ex-Norfolk & Western 2963 (formerly Illinois Terminal 2303) carries little in the way of optional equipment. On both, the turbocharger stack is visible on the roof just behind the dust evacuator dome of the inertial air filter. Photo by Louis A. Marre.

The SDL39 is one of the most interesting of the special-application locomotives developed by EMD. Milwaukee Road wanted to replace its Alco RSC2s, whose mere 237,000 pounds were spread over six axles — only 19.75 tons per axle, versus the 30 tons per axle typical of U. S. road units. The answer was the SDL39. By shortening the frame and using specially lightened export truck frames, EMD was able to hold the weight of these units to 250,000 pounds on six axles (20.8 tons per axle) and stay within the rail and bridge load limits of the branches where the RSC2s had been operating. The 10 SDL39s were built in two batches, 5 in 1969 and 5 late in 1972 after all other non-Dash 2 production had ceased.

A major saving in weight came from using the 12-cylinder turbocharged prime mover instead of the heavier 16-cylinder Roots blower version found in SD38s. Seldom is the weight of a prime mover a significant consideration in U. S. diesel locomotives, because there is usually a need to ballast locomotives for tractive effort. The only exceptions occur where it is necessary to stay within the load limits of shop cranes used for lifting out engines.

SDL39

Note the unique trucks and the short frame (just nine handrail stanchions along the long hood) which are distinguishing features of the SDL39. Soo Line, as successor to the Milwaukee Road, eventually sold the nine surviving SDL39s to Wisconsin Central. Photo by Kyle Brehm.

EMD HIGH HORSEPOWER ROAD-SWITCHERS: 645E ENGINE, 3000 H.P., C-C

Model	AC/DC	H.P.	Cyls.	Length	Truck centers	Period produced	Approximate number of units sold U. S. A.	Canada	Mexico
SD40	AC	3000	16	65'9½"[1]	40'0"	1/66-7/72	865	330	62
SD40A	AC	3000	16	70'8"	45'0"	8/69-1/70	18	None	None
SDP40	AC	3000	16	65'8"	40'0"	6/66-5/70	6	None	14
SD40-2	AC	3000	16	68'10"	43'6"	1/72-2/86	3131[2]	719[2]	107[2]
SD40T-2	AC	3000	16	68'10"	43'6"	6/74-7/80	310	None	None
SD402-SS	AC	3000	16	68'10"	43'6"	3/78-4/78	5	None	None

[1]The standard SD40/SD45 underframe was slightly revised in 1968 to this dimension from 65'8", with no change in truck centers.
[2]Production totals are through December 31, 1986.

Railroads at first were tentative about ordering SD40s as opposed to the contemporary SD45 model; they had been conditioned by the horsepower race to seek the highest horsepower per unit to maximize the benefits of unit reduction. However, when railroads experienced high maintenance expenses for a relatively minor advance in horsepower per unit, they returned to 16-cylinder models, making the SD40-2 practically the standard locomotive of the late 1970s and early 1980s.

EMD produced a few variations on the SD40 that were significant enough to merit their own model designations. The first was the

SD40

Kansas City Southern No. 632 is an SD40 with extended-range dynamic braking. The SD40 identification feature is three closely spaced radiator fans of equal size above the radiators. The otherwise-similar SD35 has two large fans flanking a small one. Photo by Louis A. Marre.

SDP40, built only for Great Northern and National Railways of Mexico to replace aging F3s and F7s in passenger service. When Amtrak was created, Burlington Northern converted the ex-GN units to freight service by removing the steam generators and thereafter they functioned the same as any other SD40.

The next variation was the SD40A, an Illinois Central special order in which SD40 machinery was used on the longer underframe of an SDP45 to accommodate a larger fuel tank. These 18 units were delivered in two orders, one in 1969 and one in 1970.

The SD40-2 succeeded the SD40 in January 1972; however, five SD40s were delivered to Detroit Edison in 1972 at its request to maintain standardization in its small fleet. The main functional change in the SD40-2 is the Dash 2 modular electronic control system and the HT-C (high traction) truck. However, Conrail purchased SD40-2s with Flexicoil trucks because of the controversy which surrounded the HT-C truck during 1976 and 1977. (Amtrak's SDP40s, which rode on the HT-C truck, rode roughly and tended to derail under certain conditions — see the section on EMD C-C cowl units for a fuller discussion.)

The first major variation in the SD40-2 was the SD40T-2, purchased only by Southern Pacific and Denver & Rio Grande Western. The variation consists of a low air intake with fans mounted beneath the radiators, which are set horizontally in the rear of the hood. This arrangement enables locomotives to gather cooler air from near track level while passing through tunnels and snowsheds on heavy grades. Rear units in consists on heavy grades tend to overheat and shut down in tunnels as units ahead of them heat the air.

The Comfort Cab or Safety Cab was introduced in 1975 on SD40-2s

SD40

Grand Trunk Western 5902 is an SD40 without dynamic braking. Photo by Louis A. Marre.

built for Canadian National. It did not rate a new model number, but the units are sometimes referred to as SD40-2(W) for "wide nose."

Finally, the SD40-2SS appeared in 1978 as a test of the Super Series electrical system in six-motor units. Burlington Northern 7049-7053 were the only such units. The units have a larger alternator but there is no external difference in the locomotive. The successor to this wheelslip control system is used in the "50" and "60" models (GP50, SD50, SD60, etc.).

Norfolk Southern 3179 is a standard SD40 built in 1971 for the Southern Railway. Besides illustrating the high short hood and long-hood-front standards of its owner, it shows clearly the front and rear "porch" effect produced by EMD's decision to standardize all SD production on the 65'8" frame needed for the 20-cylinder SD45. At the time, the SD45 was the sales leader, and it was not foreseen that the SD40 and SD40-2 would become best-selling models (5234 units; runner-up is the GP9 and GP9B, totaling 4357 units). The SD40 measures 54' over the hoods, leaving approximately 3' empty spaces between the ends of the hoods and frame ends. When the Dash 2 line was introduced in 1972, the standard six-motor frame was extended 3' more, making the "porch effect" even more pronounced on SD40-2s. Photo by Louis A. Marre.

SD40

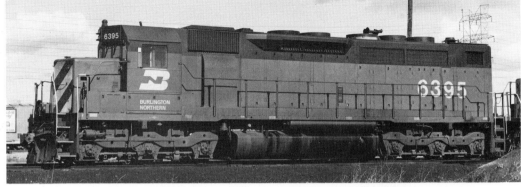

Burlington Northern No. 6395 and National Railways of Mexico No. 8524 represent the only two orders for the SDP40. All the machinery is located forward of its usual position in the SD40 to provide room for the steam generator compartment at the rear (illustrated by the small photo of BN 1976). Photo of BN 6395 by James B. Holder; photo of NdeM 8524 by Larry Russell; photo of BN 1976 by Louis A. Marre.

All SDP40

SD40A

The SD40A is the machinery of an SD40 mounted on an SDP45 frame to provide room between the trucks for a larger fuel tank. Only Illinois Central ordered the SD40A. Photo by Burdell Bulgrin.

SD40-2

CP Rail 5789 is a conventional SD40-2 equipped with dynamic braking, but without the enlarged anti-climber which has been common on SD40-2 orders since 1973. The three-lens classification lights, high bell mounting, headlight in the short hood, ditch lights, and pilot are unique Canadian features. Photo by Larry Russell.

Conrail 6396 displays the anticlimber option (the overhanging projection above the coupler), as well as the use of Flexicoil instead of HT-C trucks, unique to Conrail. Photo by Jack Armstrong.

SD40-2

Baltimore & Ohio 7602 is a typical SD40-2. It has HT-C trucks, a plain end-sill plate, screw-on battery box cover (with the venting slit preferred by Chessie System), normal-range dynamic braking (no access door ahead of the resistor grids), and the usual Dash 2 right-side-only water sight glass above the "h" in "Chessie System." Photo by Louis A. Marre.

SD40-2

SD40-2

Union Pacific 8009 is an SD40-2 having an extended nose to house radio-control equipment. Units with extended noses have also been built for Kansas City Southern and Southern Pacific. Photo by J. R. Quinn.

SP 8492 is an SD40T-2. The modified radiator extends the rear hood about 3 feet. SD40T-2s are distinguished from SD45T-2s by having two radiator fan access doors under the radiators and above the air intake, as opposed to three doors on the SD45T-2. Photo by Vic Reyna.

SD40T-2

Canadian National 5241 illustrates the Safety Cab — also called "Comfort Cab." CN has the only SD40-2s with this type of cab. Photo by Larry Russell.

SD40-2W

SD40-2

Soo Line 6618-6623, built in July 1984, were the last domestic SD40-2s delivered. Along with Soo 6617, an April 1983 rebuild of wrecked Soo 6600, they are the only domestic SD40-2s produced since Missouri Pacific 3322-3341 were delivered in January and February of 1982, reflecting how suddenly the flood of new locomotives was shut off in 1981. As a result of their late construction, the Soo units are the only domestic SD40-2s with free-flow blower ducts (left side only). Photo by Greg Sommers.

SD40-2

Canadian SD40-2 production has continued. All units since December 1982 — 61 for CP Rail and 5 for British Columbia Railway — have the free-flow blower duct. The BCR units, delivered in July 1985, stand to be the last SD40-2s built for Canada. Four SD40-2s for Mexico, NdeM 13001-13004, delivered in February 1986 (the first for Mexico since 31 units were delivered in 1980), are the last SD40-2s of all, ending a production run record of 3,957 units (4,272 counting SD40T-2 and SD40-2SS sub-models)! Photo by Larry Russell.

EMD HIGH HORSEPOWER ROAD-SWITCHERS: 645E ENGINE, 3600 AND 4200 H.P., C-C

Model	AC/DC	H.P.	Cyls.	Length	Truck centers	Period produced	Approximate number of units sold		
							U.S.A.	Canada	Mexico
SD45	AC	3600	20	65'9½"[1]	40'0"	2/65-12/71	1260	None	None
SDP45	AC	3600	20	70'8"	45'0"	5/67-8/70	52	None	None
SD45X	AC	4200	20	70'8"	45'0"	6/70-2/71	7	None	None
SD45-2	AC	3600	20	68'10"	43'6"	5/72-9/74	136	None	None
SD45T-2	AC	3600	20	68'10"	43'6"	2/72-6/75	247	None	None

[1]The standard SD40/SD45 underframe was slightly revised in 1968 to this dimension from 65'8" with no change in truck centers.

The SD45, at the top of the 645-engine horsepower line at EMD, dominated sales for the first few years after its introduction, but the cost of maintenance of the 20-cylinder engine and the big radiators soon caused railroads to question the value of the extra 600 h.p. per unit as compared with the SD40.

The Dash 2 line of 1972 addressed the radiator problem by eliminating the angular mounting. Nonetheless, most railroads had already decided. Only Santa Fe, Clinchfield, Erie Lackawanna, and Seaboard Coast Line ordered SD45-2s, while Southern Pacific and Cotton Belt accounted for all production of the "tunnel" version, the

SD45

The distinctive feature of the SD45 is its flared radiator mounting with three screened air inlets to a side. Denver & Rio Grande Western 5324 is an SD45 with extended-range dynamic braking. Photo by Kenneth M. Ardinger.

SD45T-2 (the "T" is explained in the description of the SD40T-2).

The SDP45 was developed to replace F units in long-distance passenger service, but only Great Northern and Southern Pacific bought them for that purpose. Most of those built, 34 of them, went to EL as freight power, with the long underframe used only to provide added fuel capacity. (The same underframe was used for the same reason to create the SD40A for Illinois Central.)

The SD45X was the pioneer of the Dash 2 modifications, especially the HT-C truck. It too rode on an SDP45 frame. Six SD45Xs wound up on SP's roster and the seventh was retained by EMD as a test bed. The experimental higher horsepower (4200) was not repeated, since the railroads were balking at the maintenance costs of 3600 h.p. units, as reflected in the fall-off of SD45 sales.

SD45

Only Chicago & North Western ordered SD45s without dynamic braking; No. 911 is an example. Photo by Lee Hastman.

The 20-cylinder engine and the steam generator of the SDP45 required a frame 5 feet longer than that of the SD40 and SD45. The rear view of SP 3205 shows the squared-off end housing the steam generator on the passenger version of this model, while the rear-end photo of EL 3641 shows the beveled end of the freight version. SP 3201 photo by Gordon Lloyd Jr.; photo of rear of SP 3205 by Dan Dover; photo of rear of EL 3641 by Louis A. Marre.

All SDP45

EMD 4202 exhibits the quickest identifying feature of the SD45X, four large radiator fans. Note also the HT-C trucks. Photo by Kenneth M. Ardinger.   **SD45X**

Santa Fe 5698 illustrates the primary SD45-2 identification features: the long radiator area with three widely spaced fans and the lack of long end platforms, all on an HT-C truck locomotive. These provide the easiest distinction from the SD40-2, which has the same-length frame. The box on the cab roof of AT&SF 5698 is an air conditioner. Photo by Vic Reyna.

SD45-2

Both SD45-2

In November 1972 Erie-Lackawanna acquired 13 SD45-2s with 5000-gallon fuel tanks (the normal capacity was 4400 gallons). Because of the large fuel tanks, the main air reservoirs and aftercooler piping were located in the rear of the long hood. Louvers provided ventilation for the aftercooler. (The air reservoirs and aftercooler piping are usually located in open air atop the fuel tank.) The units were intended to make a round trip without refueling from Chicago to Croxton, New Jersey, with fast intermodal trains. EL 3669-3681 became Conrail 6654-6666. Conrail 6663 photo by Gordon Lloyd Jr.; EL 3677 photo by Louis A. Marre.

SD45T-2

Southern Pacific 9204 shows the low air intake and absence of visible radiator fans of the SD45T-2. The most visible differences from the SD40T-2 are that, because of the 20-cylinder engine, the locomotive fully occupies the standard underframe without leaving the big front and back porches seen on the 16-cylinder model, and there are three radiator fan access doors on the SD45T-2 instead of two on the SD40T-2. Both photos by Louis A. Marre.

There have been occasional attempts to replace the high-maintenance 20-cylinder engines of SD45s with 16-cylinder 645-series diesels. Santa Fe completed ten such repowerings, 5426-5437, while Southern Pacific did the same for six SD45-2s, 7561-7566. In both cases no external differences are visible. All such work has now stopped, primarily because the units have passed the age where such work is economically justifiable. Normal overhauls — on ATSF this amounts to rebuild in kind — are about all that the 20-cylinder units now receive. Photo by Greg Sommers.

EMD HIGH HORSEPOWER ROAD-SWITCHERS: 645F ENGINE, 3500 AND 3600 H.P., C-C

Model	AC/DC	H.P.	Cyls.	Length	Truck centers	Period produced	Approximate number of units sold U. S. A.	Canada	Mexico
SD40X	AC	3500	16	68'10"	43'6"	9/79	4	None	None
SD50S	AC	3500	16	68'10"	43'6"	12/80	6	None	None
SD50	AC	3500	16	71'2"	45'10"	5/81-7/84	230	None	None
SD50	AC	3600	16	71'2"	45'10"	11/84-5/85	131	None	None
SD50F	AC	3600	16	71'2"	45'10"	4/85-7/87	None	60	None

Production totals are through December 31, 1987.

EMD inched forward tentatively to extract from its 16-cylinder F engine the horsepower rating that had required 20 cylinders in the E engine. The single-axle wheelslip system derived from ASEA licenses and pioneered by the SD40-2SS was an integral part of making the higher rating work. At first, the machinery was crowded onto the SD40-2 frame. The space required for electrical module racks made the arrangement crowded, and a longer frame was adopted for the production model. The SD40X, SD50S, and SD50 layouts all allowed moving the dynamic brake resistors from atop the engine to a cooler location between the central air intake and the cab.

SD40X

Kansas City Southern 700-703 of 1979 pioneered the SD50 machinery on an SD40-2 frame. There are 13 handrail stanchions instead of the 14 found on a regular SD50, because the frame is 2'4" shorter. Note the old-style blower duct. Photo by Louis A. Marre.

Both SD50S

Norfolk & Western 6500-6505, built in December 1980, were nearly identical to KCS 700-703. The "SD50" model designation had been decided upon in the meantime, so that is what they were called until the regular, production-length SD50s began to appear; then they were redesignated "SD50S" (for "short," perhaps). As built, the N&W units had the old-style blower duct, but got the free-flow design when they were returned to the factory for work in October 1981. The dynamic brake grid opening is the louver below and ahead of the central air intake. N&W 6500 photo by Louis A. Marre; 6504 photo by R. R. Harmen.

Both SD50

The first regular-length SD50s were KCS 704-713 of May and June 1981. In 1983 and 1984 the CSX roads took 144 units, including CSX 8635 (ex-Chesapeake & Ohio 8635) of October 1985. Photo by Louis A. Marre. Two units of this group, Seaboard System 8525 and 8526, had microprocessor controls in anticipation of the SD60 and EMD operated them under test conditions before delivery. In the same period Conrail took 140 SD50s. Two orders had old-style Flexicoil trucks; later orders used the HT-C design. Conrail 6712 is from the first group, 6700-6739, delivered between October and December 1983. Four latched doors in six door panels beneath the radiator grilles differentiate the SD50 from the SD60, which has six latched doors in eight door panels. Photo by Greg Sommers.

SD50+

Missouri Pacific (Union Pacific) 5000-5059, delivered in November and December 1984, are rated at 3600 h.p. They were the first uprated SD50s and were dubbed SD50+ by the locomotive enthusiast magazine *Extra 2200 South*. There is no external distinction from other SD50s, and our table assumes that all subsequent SD50s are SD50+s. Photo by George Cockle.

SD50

Denver & Rio Grande Western has the smallest SD50 fleet, 17 units built in late 1984. Photo by R. R. Harmen.

An interesting variation on the SD50 theme is the Canadian cowled version, the SD50F. The unit is rated at 3600 h.p., and could be considered an "SD50+F." Canadian National had 60 of these units at press time, Nos. 5400-5459. Behind the cab of the SD50F is the "Draper taper," a cutout in the full-width cowl that allows the enginemen some rearward vision. (It was designed by William L. Draper, CN's assistant chief of motive power and first appeared on the Bombardier HR412.) On these units the dynamic brake cooling air intakes are at the same level as the engine air intakes. Photo by Larry Russell.

SD50F

SD50 with microprocessors

Seaboard System SD50s 8525 and 8526, built in March 1984, served as testbeds for microprocessor controls in place of the usual modules and relays. Number 8525 is shown coupled to EMD's test car; the units were not lettered for Seaboard System until the tests were concluded. Photo by R. R. Harmen.

EMD HIGH HORSEPOWER ROAD-SWITCHERS: 710G ENGINE, 3800 H.P., C-C

Model	AC/DC	H.P.	Cyls.	Length	Truck centers	Period produced	Approximate number of units sold U. S. A.	Canada	Mexico
SD60	AC	3800	16	71'2"	45'10"	5/84-	365[1]	None	None
SD60F	AC	3800	16	71'2"	45'10"	10/85-	None	4[1]	None

[1]Production totals are through July 1988.

Increasing the horsepower rating of the 645F engine led to problems caused by engine stress. EMD decided to increase piston displacement and introduced the 710G engine to replace the 645F. (Piston diameter of the 710G is the same as that of the 645 engines, 9 1/16"; the stroke was increased by an inch to 11".) In addition, EMD adopted microprocessor controls.

The result was the SD60, nearly indistinguishable externally from the SD50. The single-axle wheelslip system of the SD50 and the SD60 has enabled railroads to make two-for-three substitutions for SD40-2s in heavy-haul service such as coal trains, but SD50s and SD60s have also been used in high-speed merchandise service.

SD60

The first production order of SD60s was Norfolk Southern 6554-6603, delivered from June through September 1985. NS 6596 is one of that group. Prototype units EMD1 through EMD4 and NS 6550-6553 were built in 1984, and BN 8300-8302 were built in October 1985. The major visible difference from the SD50 is six latched doors and eight door panels below the radiator intakes. Photo by Louis A. Marre.

SD60

Microprocessor controls and the new 710G engine first appeared together in demonstrators EMD1-EMD4, built between May and July 1984. Photo by R. R. Harmen.

SD60F

As with the SD50s, Canadian National chose a cowl variant, leasing four units, 9900-9903, on trial. They were delivered in October 1985 and like their SD50F predecessors have the "Draper Taper" cowl for visibility while backing, and the dynamic brake inlet at the same level as the central air intake. The builder's plates on these units refer to the model as SD50F, but the 710 engine makes the SD60F model designation preferable. Photo of 9903 by Wendell Lemon; photo of 9902 by Larry Russell.

EMD COWL UNITS: 645E ENGINE, 3000 H.P., B-B

Model	AC/DC	H.P.	Cyls.	Length	Truck centers	Period produced	Approximate number of units sold U.S.A.	Canada	Mexico
F40PH	AC	3000/3200	16	56'2"	33'0"	3/76-1/88	278	6	None
F40PH-2	AC	3000	16	56'2"	33'0"	3/85-	61	30	None
F40PH-2C	AC	3000	16	64'3"		7/87-5/88	26	None	None

Production totals are through December 1988. At press time VIA had 26 F40PH-2s on order for 1989 delivery, but it could not be determined whether F59PHs (the current model) would be substituted.

When the F40PH was introduced it was intended for short-haul and commuter trains heated by head-end electricity instead of steam. It carried a 500-kw alternator that would draw up to 710 h.p. from the prime mover for heating and lighting, and it was equipped with a 1200-gallon fuel tank. When the F40PH supplanted the SDP40F as the proposed power for Superliner trains, it was necessary to modify the F40PH with an 800-kw alternator for heating and lighting and an 1800-gallon fuel tank.

Chicago's Regional Transit Authority (RTA; it now uses the name Metra) was the first non-Amtrak customer for the F40PH. RTA spec-

F40PH

Amtrak 207 is from the first order of F40PHs (200-229). Its fuel tank is located next to the rear truck. In later units the tank is near the front truck and is larger, as on California DOT 909 and Amtrak 386 illustrated on the following pages. Photo by Jack Armstrong.

ified a 3200 h.p. rating for the units — the same as that of its F40Cs, which have the same 16-645E engine. This is simply a matter of fuelrack settings; in fact, on load test many Amtrak F40PHs are found to be set at well over 3000 h.p. When operated in multiple, only one F40PH at a time can furnish head-end power for train heating and lighting so that all 3000-plus horsepower of the other unit is available for traction.

Units built from 1985 on are designated F40PH-2 on builder's plates, but there is no substantial external difference in the models. However, Massachusetts Bay Transportation Authority 1050-1075, built in 1987 and 1988, have Cummins engine-alternator sets for head-end power to preserve full traction horsepower in single-unit operation and are designated F40PH-2C.

F40PH-2

California DOT 909, one of 18 units built between March and June 1985 for San Francisco suburban service, illustrates the later fuel tank arrangement. The unit does not have dynamic braking, as indicated by the lack of a ventilating grille at the location where it appears, by comparison, above the ''Amtrak'' lettering on the side of Amtrak 386 on the next page. (NJ Transit F40PHs also lack dynamic brakes.) Photo by Louis A. Marre.

Amtrak 386 is one of a group of 30 built in 1981 from SDP40F components and designated an F40PHR. It too has the later fuel tank. As far as is known, there is no difference between the F40PH and the F40PHR except for the trade-in credits for the latter, and no attempt is made to distinguish them in the table for this model. Photo by Stan Jackowski.

F40PHR

F40PH

Amtrak 201, from the original 1976 order, exhibits the rock screens that were soon added to F40PHs working in the East. Photo by Gordon Lloyd Jr.

F40PH-2

RTA 163, a 3200-horsepower version, as shown by its class designation, "B-32-A," illustrates that there are no external differences to go with the higher rating. This locomotive is from a 24-unit 1983 order, so it has the later underbody. Photo by Greg Sommers.

F40PH

RTA 110, from a 28-unit 1977 order, illustrates the right side of the original underbody layout. F40PH owners as of the end of 1988 are: Amtrak — 210 units (including replacements for wrecked units); Caltrans — 20; GO Transit (Toronto) — 6; Massachusetts Bay Transportation Authority — 44 (26 of them are F40PH-2Cs); Metra (Chicago) — 74; New Jersey Transit — 17; VIA Rail Canada — 30. Photo by Louis A. Marre.

GO Transit 515 is a Canadian F40PH without dynamic brakes. Photo by Bob Graham.   **F40PH**

F40PH-2

The VIA F40PH-2, as illustrated by 6402, differs from Amtrak counterparts only in optional details such as ditch lights, snowplow, headlight placement, etc. Photo by Larry Russell.

Massachusetts Bay Transportation Authority 1062 shows clearly the 8'1" extra length added to the F40 frame to make room for the Cummins diesel HEP set. Photo by Tom Nelligan.   **F40PH-2C**

EMD COWL UNITS: 645E ENGINE, 3000 AND 3600 H.P., C-C

Model	AC/DC	H.P.	Cyls.	Length	Truck centers	Period produced	Approximate number of units sold U. S. A.	Canada	Mexico
F45	AC	3600	20	67'5½"	41'8"	6/68-5/71	86	None	None
FP45	AC	3600	20	72'4"[1]	45'0"	12/67-12/68	14	None	None
SDP40F	AC	3000	16	72'4"	46'0"	6/73-8/74	150	None	None
F40C	AC	3200	16	68'10"	43'6"	3/74-5/74	15	None	None

[1]MILW FP45s were 70'8" long.

Cowl units were originally produced at the request of the Santa Fe: 9 FP45s in 1967 for passenger service and 40 F45s built a year later for freight. A cowl unit has a road-switcher frame supporting a cowl that reduces air resistance and permits troubleshooting and maintenance at speed. By contrast the body of older cab units — Es and Fs,

for example — is built like a bridge truss and is a structural component of the locomotive.

The Milwaukee Road ordered 5 FP45s, and Great Northern and successor Burlington Northern ordered 46 F45s. The FP45 was designed for conversion to freight service, and Santa Fe and Milwaukee

F45

Burlington Northern 6622 illustrates the identifying characteristic of the F45 compared to the FP45: the lack of any space between the radiator grilles and the rear of the cowl. The FP45 has space here representing the length of the steam generator platform. BN 6600-6613 were delivered as Great Northern 427-440 in 1969; 6614-6625 were ordered by GN but delivered to BN in 1970; and BN 6626-6645 were built in 1971. Photo by Gordon B. Mott.

Road converted their FP45s to freight locomotives after Amtrak began operating in May 1971.

When Amtrak decided to order its first new locomotives, the cowl unit was a logical choice. For Amtrak EMD produced the SDP40F, in effect a 16-cylinder, Dash 2 version of the FP45 riding on the Dash 2 HT-C trucks. It was a consideration that these engines could be readily converted to freight service if Amtrak failed; but Great Northern, Southern Pacific, and Santa Fe had already demonstrated that high horsepower C-Cs could replace Fs in passenger service and reduce the number of units required, so Amtrak was following accepted railroad practice. The SDP40F had a pair of skid-mounted steam generators at the rear that could be replaced easily by skid-mounted engine-alternator sets as Amtrak converted its passenger cars to head-end power.

During the same period, two Chicago-area commuter agencies which subsidized service on Milwaukee Road routes, the North Suburban Mass Transit District and the Northwest Suburban Mass Transit District, ordered 15 F40Cs (2 units for North and 13 units for Northwest). The F40C was an SDP40F with a 500-kw alternator for train lighting, heating, and air conditioning instead of the steam generators of the SDP40F. The alternator took up less space than the boilers and permitted the use of the standard SD40-2 underframe.

Soon after delivery, the SDP40Fs were involved in a number of derailments which had similar characteristics. In each (eventually about 10 accidents fit the pattern) a train was entering a curve of about 2 degrees at 40 to 60 mph when the trailing truck of the trailing unit or the lead truck of the baggage car coupled to it derailed due to a lateral force turning the rail over. Usually a small track irregularity was present, but it was well within Federal Railroad Administration (FRA) limits for Class 4 track. At the same time the SDP40F came under attack by the Brotherhood of Locomotive Engineers for heavy yaw characteristics under normal track conditions. However, little correlation could be established between the yawing and the derailments.

The matter culminated in two events: The first was a derailment on December 16, 1976, at Ralston, Nebraska, on the Burlington Northern. BN barred the SDP40F from all BN-operated Amtrak trains (they later returned under a 40-mph restriction on curves of 2 degrees or more). The other was a similar derailment 30 days later on the Louisville & Nashville at Newcastle, Alabama. The FRA issued a national 40-mph speed restriction, later lifted on roads with Class 5 track.

Amtrak decided there was something wrong with the locomotives even though extensive testing could not pinpoint anything. Unique to the SDP40F were its hollow-bolster trucks (which slightly skewed the center of rotation on the already off-center HT-C truck), its truck-center spacing, and its weight distribution. There was speculation that some kind of harmonic action was set up, either in the hollow-bolster truck alone or in conjunction with the yawing of the locomotive or the baggage car or both, that led to the destructive lateral force observed in the derailments.

Nothing was proven, but there were immediate results. Amtrak decided to trade in the SDP40Fs for F40PHs, which would have to be used even for transcontinental runs. By early 1979 more than half the SDP40Fs had been scrapped before reaching their fifth birthday. Transit districts began buying the F40PH for commuter service, leaving the F40C an orphan model. The trucks and geometry of the F40C are almost the same as those of the well-proven SD40-2, and it probably should not have been tarred with the same brush as the SDP40F.

Despite a tendency to seek somebody to blame, there is no evidence that EMD, Amtrak, or the operating railroads should have foreseen the SFP40F's problems, because the locomotive embodied what seemed to be well-engineered components with ample precedents in industry practice. What is important is that appropriate action was taken to protect the public as soon as the problem was perceived. What is unfortunate is that the research efforts could not explain the difficulty.

Milwaukee Road 3, an FP45, illustrates the space for the steam generator aft of the radiators, which distinguishes this model from the F45. Milwaukee's FP45s were numbered 1-5. Photo by Louis A. Marre.

FP45

Both SDP40F

The first 40 SDP40Fs, Amtrak 500-539, had the same pointed nose as the F45s and FP45s, as illustrated by Amtrak 521. The remaining units, 540-649, had a flattened nose which simplified sheet metal work, as shown by Amtrak 597. SDP40Fs lacked the nose platform found on F45s and FP45s, and SDP40Fs have the HT-C truck. Both photos by J. R. Quinn.

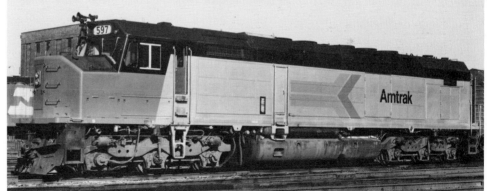

Santa Fe 5250-5267 (18 units) are SDP40Fs obtained in trade for 25 CF7s and 18 SSB1200s sent to Amtrak. The SDP40Fs were regeared for freight and equipped with front handrails at San Bernardino in 1985. They now operate in freight service along with F45s 5950-5989 (ex-5900-5939; nee 1900-1939) and converted FP45s 5990-5993 and 5995-5998 (ex-5940-5948, nee 100-108; 5944/104 has been scrapped). Photo by Greg Sommers.

SDP40F

SDP40F

EMD retained four SDP40Fs as test cars. Photo by Robert R. Harmen.

The F40C, illustrated by Metra 603, is distinguished by its corrugated side panels. Otherwise, the units have HT-C trucks and the nose style of the second-order SDP40Fs. Its 3200 h.p. rating was provided to give traction reserve when up to 700 h.p. was drawn off for train heating, lighting, and cooling. All F40Cs are operated by Metra, the rail commuter service arm of Chicago's Regional Transit Authority. Photo by Greg Sommers.

F40C

GENERAL MOTORS DIESEL (CANADA) MOTOR-GENERATOR ELECTRIC SWITCHER

Model	AC/DC	H.P.	Current	Length	Truck centers	Period produced	Approximate number of units sold		
							U. S. A.	Canada	Mexico
SW1200MG	AC/DC	1200	2300v, 60hz	44'6"	22'0"	1963-1971	None	9	None

General Motors Diesel, Ltd., predecessor of General Motors/Diesel Division, produced these motor-generator locomotives for fully automatic operation by Iron Ore Company of Canada at Labrador City, Labrador. Not only was the SW1200MG GM's first electric locomotive, it was ostensibly built on an ordinary SW1200 frame. The idea of "pulling out the engine and dropping in a rectifier" has been suggested for electric locomotive construction, and the use of standard diesel components for these units points that way. However, these locomotives narrowly preceded the universal acceptance of solid-state rectification and used the older motor-generator concept instead.

SW1200MG

Length and truck-center dimensions of the SW1200MG are the same as those of the SW1200 diesel switcher, but the unit has Blomberg trucks instead of Type A switcher or Flexicoil trucks. Photo from Peter Cox collection, courtesy of Larry Russell.

EMD/ASEA HIGH HORSEPOWER ELECTRIC BOX-CAB LOCOMOTIVES, B-B

Model	AC/DC	C.H.P.	Current	Truck Length	Period centers	produced	U. S. A.	Canada	Mexico
AEM7	AC	7000	25kv, 60hz[1]	51'1.8"	25'7.1"	11/79-	65	None	None

[1]Original current 11kv, 25hz; current changed in Northeast Corridor in October 1983.

A standard Rc4 electric passenger locomotive, demonstrator X-995, was lent by the Swedish State Railways and operated by Amtrak between New York and Washington during 1976 and 1977. The locomotive was a product of ASEA — Allmänna Svenska Elektriska Aktiebolaget, Universal Swedish Electric Co., Ltd. The locomotive forecast the general appearance of the AEM7 copies built for Amtrak under ASEA license by EMD. The Budd carbodies were built by Budd at Red Lion, Pennsylvania; electrical equipment was built and the units were assembled by EMD at La Grange.

Rc4

Noticeable differences between the Rc4 and subsequent AEM7s are lack of number indicators, different nose profile with protruding anticlimber, portholes in the side sheets, and different door placement. Photo by Donna Marvel.

102

Amtrak 911 is from the first order of 30 AEM7s, Nos. 900-929, delivered between November 1979 and August 1981. A second order of 17, 930-946, was delivered between September 1981 and July 1982, completing replacement of the GG1 electrics and displacing all but 14 E60CHs on Amtrak's Northeast Corridor route between New Haven, Connecticut, and Washington, D. C. AEM7s regularly achieve speeds of 125 mph in Metroliner service between New York and Washington. Photo by Herbert H. Harwood.

AEM7

AEM7

Maryland Department of Transportation (MARC) received the first non-Amtrak AEM7s, 4900-4903, in late 1986. Designed for slower service than the Amtrak units, they have lighter pantographs and fewer dynamic brake resistors on top. In 1987, Southeastern Pennsylvania Transportation Authority received seven similar locomotives to be used with new Bombardier Comet II coaches on push-pull trains. Both sets of carbodies were obtained from Simmering-Graz-Pauker of Vienna, Austria, because Budd — supplier of previous Amtrak carbodies — had left the railroad car business. Amtrak has two wreck-replacement AEM7s on order at press time and will presumably obtain carbodies from the same source. Photo by Herbert H. Harwood.

EMD/ASEA ELECTRIC FREIGHT LOCOMOTIVES, 6,000 H.P., C-C, AND 10,000 H.P., B-B-B

Model	AC/DC	C.H.P.	Current	Length	Truck centers	Period produced	Approximate number of units sold U. S. A.	Canada	Mexico
GM6	AC/DC	6,000	11kv, 25hz	68'10"	43'6"	4/75	1	None	None
GM10	AC/DC	10,000	11kv, 25hz	76'4"	22'0"[1]	7/76	1	None	None

[1]Distance center truck to end trucks.

The railroads began to talk about electrification again in the mid-1970s. There was another study of extending the old Pennsylvania electrification from Harrisburg west to Altoona or Conway (Pittsburgh), and Burlington Northern, Union Pacific, and Canadian Pacific studied electrification. EMD decided to protect its flank with a licensing agreement with the Swedish firm ASEA (Allmänna Svenska Elektriska Aktiebolaget — Universal Swedish Electric Co., Ltd.) for modern electric locomotive control technology and built these two demonstrators for testing on former Pennsylvania Railroad lines. However, none of the electrification studies mentioned led to a favorable conclusion, and with the decision by Conrail to terminate electric freight operations, GM6 and GM10 were returned to the builder. They were last reported still at La Grange.

GM6

The GM6 rides an SD40-2 frame. The cab sits lower than on a diesel to provide space for the pantographs. Photo by Herbert H. Harwood.

Because of its unique trucks and wheel arrangement the GM10 is on a custom underframe. Photo by Herbert H. Harwood. **GM10**

GMDD/ASEA HIGH HORSEPOWER ELECTRIC COWL LOCOMOTIVES, C-C

Model	AC/DC	C.H.P.	Current	Length	Truck centers	Period produced	Approximate number of units sold U. S. A.	Canada	Mexico
GF6C	AC	6000	50kv, 60 hz	68'10"	43'6"	11/83-7/84	None	7	None

For its new Tumbler Ridge line, extending 80 miles east from Anzac, British Columbia, to new coal mines, British Columbia Railway installed a 50,000-volt electrification using these locomotives built by General Motors-Diesel Division of London, Ontario, with ASEA-licensed electrical components. The frame length is the same as that of an SD40-2. The locomotives weigh 396,000 pounds and have HT-C trucks and E88 traction motors. The locomotives have a cab at only one end and normally operate in pairs, as with 6003 and 6005 here.

GF6C

British Columbia Railway's red-white-and-blue electrics haul unit coal trains in a remote area northeast of Prince George, British Columbia. Photo by Larry Russell.

GENERAL ELECTRIC COMPANY

General Electric was a longtime supplier of electric traction equipment to U. S. railroads. Through its role in furnishing gas-electric car transmissions, GE entered the internal-combustion rail traction field. From 1925 to 1928 GE was the electrical partner in a consortium with Ingersoll Rand and American Locomotive Company, producing 300 h.p. and 600 h.p. boxcab switchers.

In the 1930s GE developed a wide range of switching locomotives, mostly of industrial size, using a variety of diesel engines. While continuing small-locomotive production, in 1940 GE entered a marketing partnership with Alco for large road locomotives. The Alco-GE arrangement continued until 1953, when Alco resumed full responsibility for its locomotive sales. GE continued to supply electrical equipment for Alco's diesels.

During this entire period GE produced heavy electric locomotives on its own, and that business continues today.

In the 1950s GE developed its own line of diesel road locomotives for export, and in 1960 introduced to the domestic market the U25B, a 2500 h.p. road-switcher with a number of design improvements over contemporary locomotives from EMD and Alco. Within three years GE captured the No. 2 market spot from Alco and began to compete with EMD. In 1983 GE's production outstripped EMD's.

Plants: GE's heavy locomotives, both straight electric and diesel, have always been built at Erie, Pennsylvania. Some of GE's lighter locomotives were built at GE's Schenectady, New York, plant.

Engines: GE's smaller locomotives were powered by a variety of engines, primarily from Caterpillar, Cummins, and Cooper-Bessemer. The U25B used the Cooper-Bessemer FDL-16 engine. GE subsequently bought the FDL design, put the engine into production at Erie, and has used it in all its large locomotives, developing it from 2500 h.p. to 4000 h.p. in 28 years.

GE SWITCHERS: "1974 LINE," B-B, CENTER-CAB

Model	AC/DC	H.P.	Cyls.	Length	Truck centers	Period produced	Approximate number of units sold		
							U. S. A.	Canada	Mexico
SL80	DC	600	(2)6	41'0"	20'0"	11/76-	5[1]	None	None
SL110	DC	600	(2)6	44'0"	22'0"	8/74-	30[1]	None	9[1]
SL144	DC	1100	(2)6	48'0"	25'0"	12/75-	16[1]	None	13[1]

[1]Production totals are through December 31, 1986.

Although GE's 1974 Line center-cabs are primarily intended as industrial locomotives, GE demonstrated the 144-ton model as a potential Class 1 railroad switcher. Chicago & North Western leased two, Nos. 1198 and 1199, painted in full C&NW livery, for two years starting in 1979. After two years on C&NW, the demonstrators were repainted for Burlington Northern, where they spent another two years before being sold to industrial customers. The 110-ton model was ordered by industry-owned short lines — Wyandotte Southern and Moscow, Camden & St. Augustine — and the Branford Steam Railroad ordered an 85-ton SL110. Distinguishing features of the 1974 Line locomotives are the slope-sided cab and the bead outlining the hood end.

SL110

Wyandotte Southern's 110-tonner illustrates the SL110. The hood reaches practically to the stepwell. Photo by Jerry A. Pinkepank.

SL144

On C&NW's 136-ton version of the SL144 the hood is one handrail stanchion spacing back of the stepwell, because of the longer frame. On the SL80 (not illustrated) the hood extends over the stepwell a few inches. Photo by Gordon Lloyd Jr.

GE LIGHT ROAD-SWITCHER: FDL-8 ENGINE, 1800 H.P., B-B

Model	AC/DC	H.P.	Cyls.	Length	Truck centers	Period produced	Approximate number of units sold		
							U.S.A.	Canada	Mexico
U18B	DC	1800	8	54'8"	30'8"	3/73-10/76	118	None	45

Although we defined light road-switchers as locomotives having switch-engine machinery on road-engine frame and trucks, GE never built an 1800 h.p. switch engine (though it offered one). However, the U18B is clearly intended for the same market as the GP15-1, and it probably helped bring the GP15-1 into existence when Seaboard Coast Line and Maine Central traded old EMD power for U18Bs. Interestingly, U18B production ended about the time GP15-1 production began. No equivalent B18-7 appeared, nor was any Dash-8 equivalent cataloged.

When trade-in trucks were not reused, the U18B used the two-axle Floating Bolster truck seen on SCL 252. **U18B** Photo by Tom King.

A quick spotting feature for a U18B is to count the power-assembly access doors, the tall doors where "408" appears on the hood side of this Maine Central unit. A U18B has four door panels, a U23B has six, and a U30B has eight, corresponding to the number of power assemblies on each side of the engine. MEC 408 has EMD Blomberg B trucks from a trade-in. Photo by Warren Calloway.

U18B

GE MEDIUM ROAD-SWITCHERS: FDL-12 ENGINE, 2250 H.P., B-B

Model	AC/DC	H.P.	Cyls.	Length	Truck centers	Period produced	Approximate number of units sold U. S. A.	Canada	Mexico
U23B	DC	2250	12	60'2"	36'2"	8/68-6/77	425	None	40
B23-7	AC	2250	12	62'2"	36'2"	9/77-12/84	411	None	125
BQ23-7	AC	2250	12	62'2"	36'2"	10/78-1/79	10	None	None

GE waited for a year after EMD had raised its 16-cylinder engine from 2500 to 3000 h.p. (with the introduction of the 645 engine) before raising the FDL16 from 2800 to 3000 h.p. The move left GE without an intermediate-size locomotive in its catalog. There was some question whether a market for an intermediate-size locomotive existed, because up to that point railroads had been looking for higher horsepower to accomplish unit reduction when replacing 1500-1800 h.p. locomotives in mainline service.

However, EMD's GP38 began to sell well and GE decided to offer a competing 12-cylinder, 2250 h.p. locomotive, using the U30B carbody. The result was the U23B, and respectable sales followed. With the introduction of the Dash 7 line, the B23-7 succeeded the U23B.

The BQ23-7 was a B23-7 with a so-called "Quarters Cab" with extra space for a conductor's desk for operation on cabooseless trains. This cab was available on any GE Dash 7 locomotive. Because the 1978 national agreement with the United Transportation Union failed to establish a policy on the caboose and crew-consist issue that was acceptable to all carriers, individual railroads were left to negotiate their own agreements and the use of Quarters Cab locomotives will probably follow the course of local agreements.

Missouri Pacific 4667-4669, the last 3 of an order of 20 B23-7s, were delivered as 12-cylinder, 3000 h.p. B30-7As, externally indistinguishable from B23-7s.

U23B

Santa Fe 6308 has Type B trucks, which were standard on the U23B. GE's floating-bolster trucks were optional. Photo by Louis A. Marre.

Western Pacific 2258 is a U23B with EMD Blomberg B trucks from a trade-in. U23Bs can be distinguished from U30Bs by the number of power assembly access doors, the taller doors in mid-hood on which the letters ''TERN'' appear on 2258, and the letters ''ANT'' on AT&SF 6308 on the previous page. There are six door panels on U23s versus eight on U30s, reflecting the number of power assemblies on each side of the engine. Photo by Ken Douglas.

U23B

Louisville & Nashville 5129 is a standard B23-7. The door-panel means of identification (see previous page) continues in the B23-7s, though the locomotive is 2 feet longer than the U23B. The power assembly access doors are just forward of the step in the hood which identifies the Dash 7 line. Other Dash 7 features are the wide radiator and the floating-bolster truck, unless a trade-in truck is used.

This "step" in the hood, characteristic of all 7-series GE locomotives, is the result of the hood being widened to accommodate an angled, elevated position of the oil cooler. This allows the oil cooler to drain, reducing the possibility of winter freeze damage when the locomotive is shut down. Formerly the oil cooler was vertical and at floor level. Both photos by Louis A. Marre.

B23-7

Seaboard System 5139 illustrates the one-of-a-kind Quarters Cab applied to ten units of one order in 1978 and 1979 (now CSX 3000-3009). The idea of enlarging the cab to accommodate crew members when the caboose is eliminated will undoubtedly have further application. A group of BN GP50s also has a version of this feature. Photo by Greg Sommers.

BQ23-7

B23-7

Ten of the 125 Mexican B23-7s were assembled in Brazil by GE do Brasil. A further 61 units were built at NdeM's Aguascalientes shops from kits supplied by GE. The illustration shows a Brazilian-built unit, whose small fuel tank is probably intended to minimize the weight of the unit. Photo by Matt Herson, collection of Ken Ardinger.

B23-7

Conrail's 141 units ride on AAR type B road trucks, as illustrated by No. 1958, one of the 124 units (Nos. 1900-2023) built in 1978 and 1979. Conrail 2800-2816 of 1977 were the first B23-7s and the only ones built that year. Photo by Louis A. Marre.

Santa Fe 6405-6418 were the last B23-7s, and they have conventional AAR type B road trucks. They were built **B23-7** in April and May 1984 but delivery was delayed until December 1984. Photo by Greg Sommers.

As is its custom, Southern ordered B23-7s with high short hood and the long end designated as front. Photo by Warren Calloway. **B23-7**

GE HIGH HORSEPOWER ROAD-SWITCHERS: FDL-12 ENGINE, 3000 H.P., B-B

Model	AC/DC	H.P.	Cyls.	Length	Truck centers	Period produced	Approximate number of units sold U.S.A.	Canada	Mexico
B30-7A	AC	3000	12	62'2"	36'2"	6/80-2/82	58	None	None
B30-7A1	AC	3000	12	62'2"	36'2"	4/82-5/82	22	None	None
B30-7A(B)	AC	3000	12	62'2"	36'2"	6/82-10/83	120	None	None

In 1980, Missouri Pacific decided to convert the last 3 of a group of 20 B23-7s it was receiving (Nos. 4667-4669) to 3000 h.p., 12-cylinder locomotives. Compared with 16-cylinder, 3000 h.p. locomotives, fuel consumption and maintenance economies would result. Satisfied with the results, Missouri Pacific ordered 55 more such units, 4800-4854, which were delivered from January 1981 through February 1982. These locomotives are externally indistinguishable from B23-7s but as they are owned only by MP and successor Union Pacific (the units are now UP 167-169 and 200-254) the identification problem is restricted.

The next evolutionary step for this model was the B30-7A1 with the equipment blower at the cab end of the prime mover rather than the radiator end. The only examples are Southern 3500-3521.

Development of the 12-cylinder, 3000-h.p. unit concluded with two large orders of cabless versions for Burlington Northern, Nos. 4000-4052, built from June through December 1982, and 4053-4119, built between August and October 1983. They were designated B30-7A(B). The earlier units had the equipment blower at the radiator end and the later ones had it located as on the B30-7A1. One could, therefore, call these last units B30-7A1(B)s. They also had the dynamic brake resistors placed forward in a box above the equipment blower, foreshadowing the Dash-8 line.

B30-7A

Missouri Pacific 4825 rides on floating-bolster trucks and looks the same as a B23-7. Photo by Louis A. Marre.

Note the large grilled openings near the cab of 3519 in the illustration and compare it with Southern B23-7 **B30-7A1**
3971 in the preceding section. Photo by Louis A. Marre.

B30-7A(B)

BN 4017, from the first group of B30-7As, is comparable to the Missouri Pacific B30-7As except for the lack of a cab. Photo by Greg Sommers.

B30-7A(B)

BN 4083, from the final group, has the equipment blower mounted at the opposite end of the prime mover from the radiators, akin to Southern's B30-7A1s. Note also the dynamic brake grids atop the blower openings (on the opposite side of the unit, two square openings appear in this box). BN announced it was cutting GE out of its 1984 power order because it was dissatisfied with GE follow-up on service after early troubles with this group. The locomotives were subsequently reworked at Erie and have performed acceptably. Photo by Louis A. Marre.

GE HIGH HORSEPOWER ROAD-SWITCHERS: FDL-16 ENGINE, 3000-3600 H.P., B-B

Model	AC/DC	H.P.	Cyls.	Length	Truck centers	Period produced	Approximate number of units sold U. S. A.	Canada	Mexico
U30B	AC	3000	16	60'2"	36'2"	12/66-3/75	291	None	None
U33B	AC	3300	16	60'2"	36'2"	9/67-8/70	137	None	None
U36B	AC	3600	16	60'2"	36'2"	1/69-12/74	125	None	None
B30-7	AC	3000	16	62'2"	36'2"	12/77-5/81	199	None	None
B36-7	AC	3600	16	62'2"	36'2"	11/80-9/85	222	None	None

GE waited a year after EMD raised 16-cylinder horsepower from 2500 to 3000 before increasing its own 16-cylinder horsepower from 2800 to 3000. Whether this delay resulted in any significant market disadvantage is hard to say. Whatever may have been lost was probably regained when GE went to 3300 and 3600 h.p. in the 16-cylinder engine, where EMD required 20 cylinders. The U33B and U36B sold almost as many units as the U30B, though only to Auto-Train, Penn Central, Rock Island, and Seaboard Coast Line. The general decline in high-horsepower B-B sales affected these units, and most railroads were avoiding units above 3000 h.p. to save maintenance expense. Both factors are reflected in the relatively early end of production of the U33B and U36B.

U30B

Norfolk & Western 8510 illustrates the later radiator area arrangement on U30Bs. Note that there are eight power assembly door panels on the hood (they carry the letters "RFOLK AND WES"). This feature is, of course, shared by the 3300- and 3600-h.p. models which, however, have wide radiators. The high short hood on N&W 8510 is not the usual arrangement but reflects Norfolk & Western's preference. Photo by Louis A. Marre.

U28B

Norfolk & Western U28B 1900 illustrates the radiator arrangement also used on early U30Bs. The dynamic-brake grids are visible through the screening over the opening. Photo by Louis A. Marre.

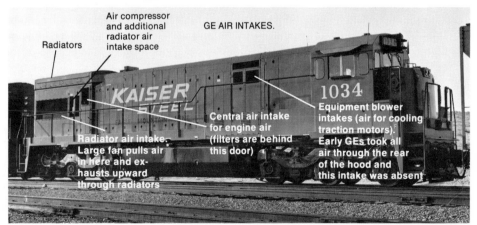

Radiators

Air compressor and additional radiator air intake space

GE AIR INTAKES.

1034

Radiator air intake. Large fan pulls air in here and exhausts upward through radiators

Central air intake for engine air (filters are behind this door)

Equipment blower intakes (air for cooling traction motors). Early GEs took all air through the rear of the hood and this intake was absent

Kaiser Steel 1034, a U30C, illustrates the hood apertures of GE U-series diesels. Photo by Kenneth M. Ardinger.

U36B

U33B

U33B and U36B units look identical. The distinction is most difficult on CSX at present: U33Bs 5600-5627 and U36Bs 5700-5805 are still in service. Most other U33B rosters have been scrapped or traded in. Rock Island 199 photo by Lee Hastman; Conrail 2973 photo by J. R. Quinn.

Frisco 864 exhibits the step in the hood (between the letters "S" and "C" in Frisco) which identifies the Dash 7 line and thereby distinguishes the B30-7 from the U36B and U33B. Any of the 3000 h.p., 3300 h.p., or 3600 h.p. GE B-Bs may be found on AAR type B, floating bolster, or traded-in EMD Blomberg B trucks. Photo by Warren Calloway.

B30-7

B30-7

From January 1980 on to comply with Environmental Protection Agency regulations, B30-7s had the fat exhaust silencer displayed by CSX 5508. B30-7s equipped thus at construction were Chesapeake & Ohio 8265-8298, Seaboard System 5500-5516, and Cotton Belt 7774-7799. As long as this feature is not retrofitted, it can become an identification feature, after excluding the above-named B30-7s, for the otherwise-identical B36-7. Photo by Stan Jackowski.

B36-7

The first B36-7s were Cotton Belt 7770-7773, illustrated by 7772. These units were originally to have been B30-7s like the 7774-7799, and their builder's plates say 3000 h.p., but they were delivered as 3600 h.p. test beds instead, in January 1980. The rest of the order came as B30-7s in March and April of that year. Photo by Paul T. Maciulewicz.

Both B36-7

After Cotton Belt 7770-7773 in January 1980, the first "production" B36-7s were Santa Fe 7484-7499 in October and November 1980. They were distinguished by sound baffles over the radiator fan area to meet EPA requirements. Presumably GE then developed a quieter fan, because later deliveries lacked this feature. Seaboard System 5826 is one of 75 delivered between February and June 1985. Eight power assembly access doors to a side tell the story of a 16-cylinder engine. If a unit otherwise looks like a B30-7 but has the big silencer stack, it is a B36-7 unless it is one of the 77 C&O, SBD, or SSW units enumerated on the previous page, or unless silencer stacks begin to be retrofitted. So far EPA regulations do not require that. Both photos by Greg Sommers.

GE HIGH HORSEPOWER ROAD-SWITCHER: FDL ENGINE, 3150, 3800, and 3900 H.P., B-B

Model	AC/DC	H.P.	Cyls.	Length	Truck centers	Period produced	Approximate number of units sold U. S. A.	Canada	Mexico
B32-8	AC	3150	12	63'7"	36'7"	1/84	3	None	None
B36-8	AC	3800	16		39'4"	10/82	1	None	None
B39-8	AC	3900	16	66'4"	40'1½"	1/84-2/88	143	None	None
Dash 8-40B	AC	4000	16	66'4"	40'1½"	5/88-	89	None	None

The first of the Dash 8 line of microprocessor-controlled GE locomotives was test unit GE 606, the B36-8, a locomotive initially rated at 3600 h.p. and then run up to the higher rating shown. This model was not cataloged, and No. 606 has since been rebuilt twice.

Next came GE-owned B32-8s 5497-5499, which have been on long-term test on Burlington Northern. These were originally to have been B30-8s but were slightly uprated to 3150 h.p.

During 1984 GE fielded three more test units, B39-8s in Santa Fe colors, ATSF 7400-7402. All of these units were considered test beds rather than cataloged production models.

B36-8

There was only one B36-8, demonstrator No. 606. It was later rebuilt to become B39-8 demonstrator 808. General Electric photo.

B32-8

Burlington Northern 5499, actually one of three GE demonstrators extensively tested on BN, exhibits the main Dash 8 identification features: tapered nose, equipment blower and dynamic brake grid box above the hood roof line behind cab, and squared radiator overhang. The unit load-tests at 3000 h.p. with all parasitic loads accounted for. The higher rating is based on the microprocessor control's ability to minimize the effect of parasitic load on tractive effort. Photo by Greg Sommers.

B39-8

Both sides of B39-8 prototype Santa Fe 7400 (actually a GE demonstrator) illustrate the difference in hood openings on the left and right sides of the Dash 8 prototypes. Both photos by Greg Sommers.

Production B39-8 units differ only slightly from ATSF 7400. The radiator air intake grilles are angled rather than vertical, and the intake openings beneath them have been regularized and reduced in number. Following delivery of only 140 production units, GE announced a change of nomenclature and a slight horsepower increase for future versions of this model: Dash 8-40B. B39-8 production comprised 40 units for Southern Pacific (8000-8039) and 100 for LMX Leasing (8500-8599), plus two of those rebuilt after wrecks but not relisted. Photo by Gordon Lloyd Jr.

B39-8

B39-8(W)

In October 1988 both GE and EMD announced "Canadian cab" designs applicable to all cataloged road units. GE was the first to field a prototype, former B39-8 test unit 808 renumbered 809 when the new cab was applied. Not merely an external modification, the new cab incorporates a console control display, the first substantial departure from the control stand cab design which evolved from electric locomotive cab layout during the infancy of diesel-electric design. Provision for air conditioning and space for the entire train crew are also features of the new cab. Both builders anticipate that this cab will become the standard for future orders. If it does, it will certainly alter the looks of American locomotives through the 1990s. Both photos by George Cockle.

GE HIGH HORSEPOWER ROAD-SWITCHERS: FDL-16 ENGINE, 3000-3600 H.P., C-C

Model	AC/DC	H.P.	Cyls.	Length	Truck centers	Period produced	Approximate number of units sold U. S. A.	Canada	Mexico
U30C	AC	3000	16	67'3"	40'11"	1/67-9/76	592	None	8
U33C	AC	3300	16	67'3"	40'11"	1/68-1/75	375	None	None
U34CH	AC	3430	16	67'3"	40'11"	11/70-1/73	32	None	None
U36C	AC	3600	16	67'3"	40'11"	10/71-4/75	124	None	94
U36CG	AC	3600	16	67'3"	40'11"	4/74-5/74	None	None	20
C30-7	AC	3000	16	67'3"	40'11"	9/76-2/85	783	None	354
C36-7	AC	3600[1]	16	67'3"	40'11"	6/78-12/85	129[1]	None	40

[1]85 U. S. units built from June 1985 on were rated at 3750 h.p.

The U30C was an evolutionary rerating of the U28C, and the first U30Cs were visually indistinguishable from late U28Cs. The U33C and U36C were essentially the same locomotive as the U30C but with advanced fuel rack and greater radiator area. In addition, the U36C had steel-capped pistons. The ability to develop 3600 h.p. with 16 cylinders gave GE a sales advantage for a brief season, as compared to EMD's 20-cylinder locomotives, but this advantage melted away as railroads concluded then that locomotives over 3000 h.p. were not worth the extra expense of maintaining them.

The U34CH is a U36C fitted with a train-lighting alternator, giv-

U30C

Kaiser Steel 1030 represents the first style of U30C carbody after the U28C style, with the radiator end of the hood wider than the rest and a small fairing piece at the top of the hood at radiator level. (This fairing is absent from Rock Island 4598 on the following page.) Note that on the Kaiser Steel unit the forward-most air intake on the radiator bulge is high, whereas on the RI unit it is low. These position variations originally depended on the type of air filter used; later it was no longer a consistent indication. Photo by Joe McMillan.

ing a nominal rating of 3430 h.p., although this varies downward depending on the lighting demand of the train. At one time there was a plan to have GE buy back a C&NW U30C to convert to a thirty-third U34CH, because the changed layout of the 7-series locomotives does not leave room for a train-lighting alternator (which occupied the "steam generator room" behind the cab on the 6-motor U-series engines). The U36CG was simply a case of putting a steam generator in this compartment.

Note the absence of the fairing at the top of the hood forward of the radiator (characteristic of units built after September 1978) and the low air intake at the front of the radiator compartment. Photo by Louis A. Marre.

U30C

The only difference in appearance between the U30C and the U33C, exemplified by Burlington Northern 5761, is the wider radiator area. The numerals "61" and the ACI label on the side of the hood extend across the eight power assembly access door panels. Eight panels indicate a 16-cylinder engine; 12-cylinder models have six of these high door panels. The lack of a wider hood and no "step" between "57" and "61," distinguishes this from the C30-7. Photo by Kenneth M. Ardinger.

U33C

U34CH

NJ Transit 4171 illustrates the U34CH, which preceded U36C production by a year, although the engine used the steel-capped pistons of the U36. The train lighting alternator occupies the space behind the cab. Photo by Jim Herold.

C30-7

The final 148 of NdeM's 328 C30-7s, Nos. 11000-11148, were assembled from GE kits at Aguascalientes shops. Not having to meet U. S. noise standards, these units do not have the exhaust silencer found on U. S. C30-7s built since January 1980. Photo by Kenneth M. Ardinger.

BN 5583 illustrates the identifying feature of the C30-7, a hood that widens slightly about two-thirds of the way back, just aft of the power assembly doors. The increase in width is stepped rather than tapered or faired; you can see it between the numerals "55" and "83." The 3000h.p. C30-7s have the wide radiator area previously identified with 3300 and 3600 h.p. units. Although C30-7 production has otherwise ended, GE has 20 unfinished NdeM Aguascalientes kits at Erie which cannot be delivered due to Mexico's international monetary problems and may wind up being assembled and delivered domestically. Photo by Greg Sommers.

C30-7

U36C

Santa Fe 8700 is a U36C, visually indistinguishable from the U33C. The only U36Cs were built as AT&SF 8700-8799; Clinchfield 3600-3606; Erie Lackawanna 3316-3328; Milwaukee Road 8500-8503; National of Mexico 8900-8937, 8958-8986, and 9300-9316; and Ferrocarril del Pacifico 409-418. Photo by J. R. Quinn.

U36CG

NdeM 8945 is a U36CG, one of 20 numbered 8938-8957. The U36CG is a U36C with a steam generator occupying the compartment provided for this purpose on all U-series C-Cs. Note the housing for steam generator vents on the roof behind the cab. Photo by Keith E. Ardinger.

The first C36-7 was an engine department test unit numbered GE 505, built in June 1978. The first production models were for Mexico: National Railways of Mexico 9317-9326 in March 1979, followed by Ferrocarril del Pacifico 419-433 between July and October 1979. (A subsequent Pacifico group, 434-455, was ordered as C36-7s but delivered as C30-7s.) These units were then followed by NdeM 9327-9341 between March and May 1980. All NdeM's C36-7s were built in Brazil. The first domestic C36-7s were Norfolk & Western 8500-8505 of March 1981, followed by 8506-8530 in May and June 1982. Photo by Louis A. Marre.

C36-7

C36-7 with Dash 8 components

No C36-7s were built in 1983. When units were again produced under that model number, they had Dash 8 features, primarily dynamic brake grids in a high box over an enlarged equipment blower fan at the cab end of the prime mover. The horsepower rating was raised to 3750, but the first units to appear in this configuration, Norfolk Southern 8531-8542 of May 1984, were still rated at 3600 h.p. Photo by Kenneth M. Ardinger.

Both C36-7

Conrail 6620-6644 of June 1985 were the first units for which the 3750 h.p. rating was announced, but they were throwbacks in appearance to earlier C36-7s. Union Pacific (Missouri Pacific) 9000-9059 built from October to December 1985 finished out the model. These last three groups, 97 out of the 129 C36-7s in the U. S., could almost be designated "C38-7.5." The photos of 9006 and 9047 show the differences on the two sides of this configuration of C36-7. Both photos by Louis A. Marre.

Conrail C36-7s ("C38-7.5s") 6620-6644 were the first rated at 3750 h.p., but they lacked the Dash 8 appearance features of the NS engines which preceded them. Photo by Louis A. Marre.   **C36-7**

C36-7

C36-7 prototype GE 505 was built in June 1978 as an Engineering Department test unit. By June 12, 1985, when these photos were taken at Council Bluffs, Iowa, the unit had been reworked with Dash 8 components and resembled the 1984-85 NS and UP units. Both photos by George Cockle.

GE COWL UNITS: FDL-16 ENGINE, 3000 H.P., C-C

Model	AC/DC	H.P.	Cyls.	Length	Truck centers	Period produced	Approximate number of units sold U. S. A.	Canada	Mexico
U30CG	AC	3000	16	67'3"	40'11"	11/67	6	None	None
P30CH	AC	3000	16	72'4"	46'0"	8/75-1/76	25	None	None

Santa Fe ordered passenger units with full-width bodies from GE as well as from EMD — 6 U30CGs that were delivered at the same time as the 9 FP45s. The GE unit was built on the same frame as its standard 3000 h.p. C-Cs, because GE's C-C design had originally allowed space for a steam generator right behind the cab. (Only 30 GE C-C hood units were built with steam generators: 10 U28CGs for Santa Fe and 20 U36CGs for National Railways of Mexico.)

The P30CH, built for Amtrak, had an auxiliary engine and generator at the rear of the locomotive to furnish power for light, heat, and air conditioning. The "Pooches" are generally assigned to the New Orleans-Los Angeles *Sunset Limited* and the Lorton, Virginia-Sanford, Florida *Auto Train*.

U30CG

Santa Fe's U30CGs were delivered painted red and silver for passenger service. Photo by Jerry F. Porter.

U30CG

Santa Fe assigned the units to freight service in 1969 and repainted them blue and yellow. After Amtrak began operation in May 1971, Santa Fe removed the steam generators. The units were traded in on an order of B36-7s in 1980. Both photos by Louis A. Marre.

Amtrak 724, highest numbered of the series, illustrates the front and side of the P30CH. The central air intakes are aft of the radiator intakes. Photo by Gordon B. Mott.

P30CH

On the rear of the P30CH are louvers for ventilating the auxiliary engine. Photo by J. R. Quinn.

Amtrak finds its remaining 6-motor diesels, the P30CHs, most useful on the long, heavy *Auto Train*. Because the units have separate engine-alternator sets to provide "hotel" power for the cars, the full 3000 h.p. of each locomotive's prime mover is available for traction. Photo by Alex Mayes.

GE HIGH HORSEPOWER ROAD-SWITCHER: FDL-12 ENGINE, 3000 and 3150 H.P., C-C

Model	AC/DC	H.P.	Cyls.	Length	Truck centers	Period produced	Approximate number of units sold		
							U. S. A.	Canada	Mexico
C30-7A	AC	3000	12	67'3"	40'11"	5/84-6/84	50	None	None
C32-8	AC	3150	12	67'11"	40'7"	9/84-	10	None	None

Production totals are as of December 31, 1988.

After the last domestic 16-cylinder C30-7s were built in 1982, all 1983 passed with no U. S. deliveries. In the meantime, the 12-cylinder, 3000 h.p. locomotive had become a GE staple, with 200 B30-7As delivered from 1980 to 1983. In 1984 GE combined the 12-cylinder, 3000 h.p. engine with 6-axle running gear to create the C30-7A. A single order of 50 units for Conrail, 6550-6599, turned out to be the only C30-7As built; just three months later the first C32-8s were delivered — Conrail 6610-6619.

C30-7A

There was no "C23-7" to be upgraded to a C30-7A, but that in effect is what the C30-7A looks like — note there are six power assembly access doors instead of eight in the illustration of Conrail 6582. Photo by Greg Sommers.

Conrail 6618 displays the unmistakable identification features of the Dash 8 line — beveled nose, dynamic brake and equipment blower hump next to the cab, and squared-off radiator. Six power assembly access doors distinguish the C32-8 from the C39-8, which is longer and has eight such doors. Photo by Jack Armstrong.

C32-8

GE HIGH HORSEPOWER ROAD-SWITCHER, FDL-16 ENGINE, 3900 and 4000 H.P., C-C

Model	AC/DC	H.P.	Cyls.	Length	Truck centers	Period produced	Approximate number of units sold U. S. A.	Canada	Mexico
C39-8	AC	3900[1]	16	70'8"	43'4"	3/83-12/87	162	None	None
Dash 8-40C	AC	4000	16	70'8"	43'4"	12/87-	100[2]	30[2]	None

[2]Production totals are as of December 31, 1988.
[1]Prototype unit was 3600 h.p.

After trials with test unit GE 607, which was uprated from 3600 to 3900 h.p., two more prototypes were sent out in January 1984, lettered Norfolk Southern 8550 and 8551. The first production models were NS 8552-8563, delivered in October and November of 1984. The Dash 8-40C replaced the C39-8 in GE's line at the end of 1987.

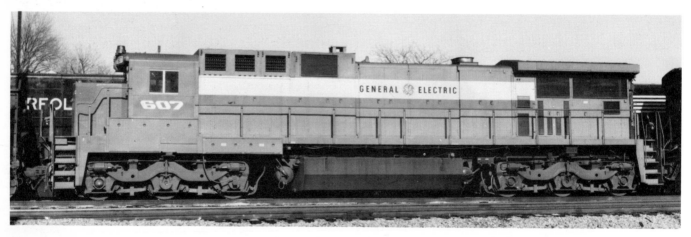

Eight power assembly doors per side distinguish the 16-cylinder C39-8 from the 12-cylinder C32-8, which has six doors on each side. Photo by Bob Graham.   **C39-8**

C39-8

Conrail 6008 and 6014 illustrate the different hood openings on the right and left sides of Dash 8s. Closeup views show the hood openings in the equipment blower and dynamic brake hump next to the cab. The dynamic brake grids are nearest the cab. Air is drawn in by fans from the right side and exhausted over the grids on the left. The larger opening farther from the cab is for the equipment blower, which takes in air on the right and exhausts it on the left, like the dynamic brake blower. The main alternator and rectifier are located beneath this blower, as well as the contactors for traction and dynamic braking current. All photos by Louis A. Marre.

The Dash 8 represents a considerable departure in styling for GE, perhaps intended to convey an impression of a no-nonsense machine (not all will agree). Conrail 6014 shows the angles of the short hood, while NS 8611 shows the beveling of the long hood (which, as usual, is front on NS units). Conrail 6014 photo by Louis A. Marre; NS 8611 photo by Greg Sommers.

C39-8

The first 4000 h.p. production locomotives delivered with the model designation "Dash 8-40C" were 75 units for Union Pacific. Delivery began in December 1987. As with production B39-8s, the radiator air intakes are canted outward, somewhat like SD45 radiators, and the carbody air intakes and vents have been tidied up from the test unit arrangements. This is likely to be the appearance of the hoods of GE's diesels for many years, as most major mechanical and electrical changes for the "Dash 8" series are now in place after a period of development which began with test unit GE 505 in 1978. Union Pacific's subsequent Dash 8-40C orders will include the "Comfort Cab," but the hood styling will probably remain the same. Photo by George Cockle.

Dash 8-40C

GE HIGH HORSEPOWER ELECTRIC ROAD-SWITCHERS, 2500 H.P., B-B; 5000 H.P., C-C

Model	AC/DC	C.H.P.	Current	Length	Truck centers	Period produced	Approximate number of units sold U. S. A.	Canada	Mexico
E25B	AC/DC	2500	25kv, 60hz	61′2″	37′10″	5/76-2/79	7	None	None
E44	AC/DC	5000	11kv, 25hz[1]	69′6″	45′0″	12/60-7/63	66	None	None
E50C	AC/DC	5000	25kv, 60hz	69′6″	45′0″	5/68	2	None	None

[1]Survivors are to be converted to 25kv, 60hz.

Texas lignite coal has marginal heating properties, attested to by the continued importation to Texas of Wyoming coal despite acrimonious complaints about freight rates. Nevertheless, Texas Utilities, Inc., has power plants which burn lignite. They are served by mine-to-plant electrified railroads.

Texas Utilities' first five electric locomotives were delivered in May 1976: Nos. 2304, 2305, 2306 (numbered just beyond the company's diesels), and 3301 and 3302 (which were like the first three, de-spite their different number series). Two more were delivered in December 1978 (3303) and February 1979 (3304).

The locomotives are employed as follows: Nos. 2304-2306 work on a 10-mile railroad at Monticello station, near Mt. Pleasant, Texas. They were built with U23Bs 2301-2303, which now haul the fly ash train there. Numbers 3303 and 3304 operate on a 13-mile railroad at Martin Lake, near Henderson. Both trains are one-man, push-pull operations with a control flat car at the opposite end from the locomo-

E25B

Texas Utilities No. 2305 is shown at the Monticello power station. The unit is slightly longer than GE's contemporary U23B. Photo by Bryan Griebenow.

tive. Trains shuttle between the mine silo and the power plant dump. At each end of the run the operator gets off to run the train through the loading silo or the dumping area by remote control (hence the remote-control lights mounted on a handrail stanchion).

In 1982 Texas Utilities announced plans to build a third such railroad, 20 miles long, at the Twin Oaks generating station near Bremond. Locomotives have not yet been ordered for this line.

E25B

The left-side view of No. 3301 shows hood openings almost the same, but a protective panel shields a portion of the rooftop switchgear on this side and not on the right. The five different-colored lights mounted on a handrail stanchion on each side show the remote operator the control status of the locomotive. Photo by Bryan Griebenow.

In the late 1950s the Pennsylvania Railroad sought replacements for its P5a electrics. Pennsy adopted an uprated version of the 3300-c.h.p., C-C, rectifier freight locomotives which GE built for the Virginian Railway in 1956-57. (Norfolk & Western sold Virginian's rectifier electrics to the New Haven; they later became Conrail class E33.) Pennsylvania acquired 66 such locomotives, designated E44, between 1960 and 1963. They were originally rated at 4400 h.p. and were rerated to 5000 c.h.p. when their mercury-ignitron rectifiers were replaced with solid-state silicon rectifiers in the 1970s. The new classification was E44A.

In the early 1980s Amtrak was planning to convert its electrification from 11kv, 50hz to 25kv, 60hz. Conrail sent E44 No. 4453 to GE's Erie shops in 1981 for conversion to the new current, and it was held there until the change so it could be tested, then serve as a prototype for converting the rest of the E44 fleet. It came from Erie in March 1984 and was tested on TOFC trains between Potomac Yard and Meadows. However, the need to convert all its E44s and disputes over the cost of sharing electrification facilities with Amtrak, SEPTA, Maryland DOT, and New Jersey Transit led to Conrail's decision to drop freight electrification and, in June 1984, retire all the E44s.

At that time, New Jersey Transit had eight E44s, Nos. 4458-4465, on hand as potential GG1 replacements, but none had been altered for the new current. NJT decided not to convert its E44s and in December 1986 sold them to Amtrak. Amtrak took them to Wilmington for conversion and plans to use them for work trains. The remaining Conrail E44s were traded in to GE on new diesels.

E44

The major difference in appearance between the E44, exemplified by Conrail 4463, and contemporary diesel hood units is a lower cab with a pair of pantographs mounted atop it. Photo by Warren Calloway.

E44

Amtrak 502 has new silver and black paint after its conversion to 25kv, 60hz current, but it still displays its Pennsylvania-Penn Central-Conrail number next to the headlight. Photo by Herbert H. Harwood.

E50C

The E50C locomotives are near-duplicates of the E44. They were built for the Muskingum Electric Railroad, an Ohio mine-to-power-plant shuttle operation akin to those of Texas Utilities. Such lines are becoming the last bastion of freight electrification in the United States. These units anticipated by 13 years the conversion of the E44s to 25kv, 60hz, as this standard commercial current was used from the outset at Muskingum. Photo by John B. Corns.

GE COWL-BODY ELECTRIC FREIGHT AND PASSENGER LOCOMOTIVES, 6000 H.P., C-C

Model	AC/DC	C.H.P.	Current	Length	Truck centers	Period produced	Approximate number of units sold U. S. A.	Canada	Mexico
E60C	AC	6000	50kv, 60hz	63'2"	36'10"	12/72-10/76	6	None	None
E60CH	AC	6000	25kv, 60hz[1]	71'3"	45'4"	10/74-8/75	26	None	None
E60C-2	AC	6000	25/50kv, 60hz			8/82-1/83	2	None	39

[1]Original current was 11kv, 25hz; Northeast Corridor converted to 25kv, 60hz in October 1983; 2 units sold to Navajo Mine subsequently converted to 50kv, 60hz.

GE first used a cowl-type body on a contemporary electric locomotive when the company built six E60C locomotives for the Black Mesa & Lake Powell. The sealed, box-like design had obvious advantages in the blowing dust of the Arizona desert where this isolated mine-to-power-plant operation is located.

Amtrak's 7 E60CPs and 19 E60CHs of 1974 and 1975 were GE's next cowl-body electrics. The E60CPs had steam generators for the remaining steam-heated equipment operating in the Northeast Corridor, while the E60CHs were equipped to provide head-end electricity for Amfleet coaches. They were equipped for 11kv, 25hz current

E60C

Black Mesa & Lake Powell E60Cs are distinguished from the E60CHs and E60C-2s by having a cab at only one end. Photo by Kenneth M. Ardinger.

but were designed for easy conversion to 25kv, 60hz current (the Corridor was converted in October 1983).

Tracking problems (which coincided with similar problems with the SDP40F diesels) held up delivery of the units. Amtrak became disenchanted with the E60s, and the Federal Railroad Administration limited them to 90 mph in a corridor where Amtrak wanted to operate at 120 mph. Amtrak's plan to phase out the E60s as AEM7s

were delivered and sell them to Conrail as freight locomotives died with Conrail's decision to cease using electric locomotives. In October 1982 E60CHs 966 and 968 were sold to the Navajo Mine for coal trains, and in September 1983 Nos. 958-963, 967, and 971-973 were sold to New Jersey Transit. The remaining units have been retained by Amtrak and renumbered 600-613. The seven of those that were E60CPs (formerly 950-956) have been converted to E60CHs.

E60CP

The E60CP was distinguished from the E60CH by the roof appurtenances for the steam generator visible just behind the cab roof opposite the pantograph end. Amtrak 950 photos by Gordon B. Mott. The E60CH, such as Amtrak 968, has just a roof vent for the head-end power alternator at this position and battery boxes instead of a water tank between the trucks. Photo by Herbert H. Harwood.

E60CH

Two Amtrak E60CHs, 966 and 968, were sold in October 1982 to Navajo Mine for a mine-to-power-plant operation similar to the BM&LP, and were repainted and renumbered LOE 20 and LOE 21. The new pantograph
mounts reflect increased catenary height. The railroad operates between Fruitland and Farmington, New Mexico. Photo by Brian Griebenow.

E60CH

The latest version of the GE cowl electric is the E60C-2, as built for Western Fuel's Deseret-Western Railway, which runs between Deserado mine near Rangeley, Colorado, and Deseret Power Station near Bonanza, Utah, 34.5 miles; and for National Railways of Mexico. The 39 Mexican units are sitting under plastic wrappings in Mexico because the country's economic problems have not allowed completion of the 212-mile, double-track electrification between Mexico City and Irapuato. (The first phase, Mexico City to Queretaro, about 145 miles, was completed in early 1983 but is probably too short to operate economically.) Both photos by General Electric.

E60C-2

MONTREAL SUBURBAN SERVICE ELECTRIC LOCOMOTIVES

Type	Builder	AC/DC	C.H.P.	Current	Length	Truck centers	Period produced	Number of units Canada
Boxcab	GE	DC	1100	2400v	37'4"	articulated	6/14-11/16	6
Boxcab	EE[1]	DC	1100	2400v	40'0"	articulated	1924-1926	9
Center-cab	GE	DC	1100	2400v	42'9.9"	21'0"	7/50	3

[1]English Electric

The GE boxcabs are the original locomotives of Canadian Northern's Mount Royal Tunnel electrification. Canadian National, successor to Canadian Northern, bought the English Electric boxcabs from the National Harbours Board in the early 1940s to increase the capacity of the commuter line. The GE center-cab locomotives strongly resemble center-cab diesels and were purchased to supplement the boxcab fleet as traffic grew. These electrics, plus six M.U.-car sets consisting of a motor car and two trailers built in 1952, were still running in Canadian National colors as of May 1988, although they have been operating for Montreal Urban Community Transportation Commission (MUCTC) since 1982.

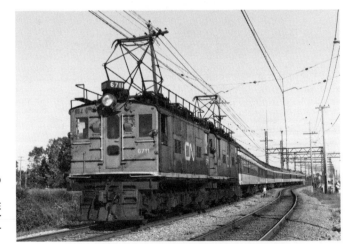

GE boxcab

More advanced in years than many of the passengers they haul, the GE boxcabs are the oldest non-preserved, non-museum locomotives in regular daily service in North America. They are numbered 6710-6715. Photo by Gordon Lloyd Jr.

CN's center-cab electrics, 6725-6727, resemble GE's 80-ton diesel of the late 1940s and early 1950s with the addition of pantographs. Photo by Gordon Lloyd Jr.

GE center-cab

The English Electric boxcabs have a narrower carbody with protruding cab sides at one end. They carry numbers 6716-6724. Photo by Gordon Lloyd Jr.

English Electric

MONTREAL LOCOMOTIVE WORKS
BOMBARDIER INC.

Montreal Locomotive Works began its corporate life as a Canadian branch of the American Locomotive Company (Alco). After Alco left the locomotive business in 1969, MLW took over and continued producing Alco's designs, doing a modest amount of business constructing locomotives for export and what little remained of the Canadian locomotive market. MLW also licensed Alco designs for foreign assembly.

In 1979 MLW was acquired by Bombardier Inc., a general manufacturing firm best known for skimobiles. Bombardier refined some of Alco's designs and marketed them without much success, several times announcing its intention to withdraw from the heavy locomotive market. The lightweight LRC, a final attempt to create a new market, was not altogether successful, and at this writing Bombardier is not producing locomotives for North American customers.

MLW SWITCHERS: 251C ENGINE, 2000 H.P., B-B

Model	AC/DC	H.P.	Cyls.	Length	Truck centers	Period produced	Approximate number of units sold		
							U.S.A.	Canada	Mexico
M420TR	DC	2000	12	50′0″	25′4″	4/72	None	2	None
M420TR-2	DC	2000	12	57′2″	32′2″	7/75	None	None	15

Using the same machinery as in the M420 road locomotives, the orders for this model to date — 2 M40TRs for Alcan, Ltd.'s, Roberval & Saguenay, and 15 M420TR-2s for Mexico's Ferrocarril del Pacifico — may be regarded as custom-built jobs, tailored to the needs of a particular customer. Presumably, most potential customers found little price advantage in foregoing the flexibility of a conventional M420 to buy one of these machines.

Only two of this model were built, Roberval & Saguenay Nos. 26 and 27. Dave More photo, from collection of Larry Russell. **M420TR**

M420TR Type 2

Only one order of this model was built, 15 units for Mexico's Ferrocarril del Pacifico. The cab end, with its distinctive "eyebrow" holding the number indicators, headlight, and classification lights, is considered the front of these units. The "1800 HP" barely visible under the number on the cab indicates derating by the railroad. Both photos by Paul C. Hunnell.

MLW MEDIUM ROAD-SWITCHERS: 251C ENGINE, 2000 H.P., B-B

Model	AC/DC	H.P.	Cyls.	Length	Truck centers	Period produced	Approximate number of units sold U. S. A.	Canada	Mexico
M420	AC	2000	12	60'10"	36'5"	5/73-2/77	None	92	None
M420B	AC	2000	12	60'10"	36'5"	6/75-7/75	None	8	None
M420R	AC	2000	12	60'10"	36'5"	2/74-5/75	5	None	None
HR412	AC	2000	12	60'10"	36'5"	9/81-11/81	None	11	None

When Montreal Locomotive Works was building Alco designs under license, no equivalent of the Century 420 was sold in Canada. Instead, MLW continued building the 1800 h.p. RS18 until June 1968, and also built 92 Century 424s through May 1967 for Canadian National and Canadian Pacific. In 1969, MLW purchased Alco's engineering designs and took over Alco's worldwide locomotive licensing agreements. MLW replaced the Century line with its own "M" line of Montreal-designed power, which was built around the same 251 engine and GE electrical equipment that had characterized the Alco-licensed production.

Between 1969 and 1972, MLW built M630s and M636s for domestic markets (plus the experimental M640), but spent 2 years on a 2000 h.p. design for the Roberval & Saguenay, finally delivered in 1972 as the M420TR. The M420 followed shortly afterward, but as an AC machine with the "Comfort Cab," which had just come into use in Canada that year.

The M420B was a cabless version designed to be used as a radio-controlled mid-train helper for the British Columbia Railway, while the M420R was designed to use components from traded-in Alco RS-3s (mainly trucks) for the Providence & Worcester.

M420

British Columbia Railway 645 has a full-width nose and a Comfort Cab. The unit has dynamic braking, evidenced by the openings at the top of the hood behind the central engine air intake. The middle square in the dynamic brake grouping is a bare, unprotected dynamic brake resistor grid. It is unusual that this is not protected by screening because of potential heat and shock hazards during dynamic braking. Compare it with the photo of Canadian National 2505. Photo by Larry Russell.

Canadian National 2505 lacks dynamic braking. The truck used on the M420 and M420B is a high-adhesion design known as the Zero Weight Transfer or ZWT truck. Photo by Larry Russell. **M420**

M420B

This is what would normally be the cab end of the unit, evidenced by the position of the central air intake, dynamic brake cooling openings, and radiator. Radio control equipment is housed in the "cab" end of the hood. Photo by Larry Russell.

M420R

Providence & Worcester 2004 is similar to the standard M420 except for Alco trade-in trucks. MLW 420s are distinguished from other Comfort Cab locomotives by the bevel along the top edge of the hood. Photo by Ken Douglas.

Canadian National 2582 exemplifies the HR412 variant of the M420 design. The main visible difference on these units (CNR 2580-2589 built between September and November 1981 and Bombardier demonstrator 7000 of September 1982) is the arrangement of the radiator area. Although CNR took its HR412s at a 2000 h.p. rating, they were cataloged at 2400 h.p. (or at least, as 2000 h.p. machines with potential to be upgraded to 2400 h.p.). Photo by Gordon Lloyd Jr.

HR412

MLW HIGH HORSEPOWER ROAD-SWITCHERS: 251E ENGINE, 3000 AND 3600 H.P., C-C

Model	AC/DC	H.P.	Cyls.	Length	Truck centers	Period produced	Approximate number of units sold U. S. A.	Canada	Mexico
M630	AC	3000	16	69'6"	41'10"	11/69-11/73	None	55	20
M636	AC	3600	16	69'6"	41'10"	11/69-4/75	None	95	16
M640	AC	4000	18	69'6"	41'10"	2/71	None	1	None
HR616	AC	3000	16	69'6"	41'10"	2/82-8/82	None	20	None

Models M630 and M636 are Montreal Locomotive Works versions of the Alco Century 630 and Century 636 with slight engineering changes. The HR616 is a full-width-body version which MLW adver- tised at 3200 h.p., but only Canadian National 2100-2103, while leased back to Bombardier to work as demonstrators on CP Rail, ran at this rating.

M630

The M630 and M636 differ from the Alco Century C630 and C636 in having a faired aftercooler box which appears at the top of the hood behind the cab. This box is not faired on the Centuries. The M630 is distinguished from the M636 by the fairing where the radiator joins the long hood. The M636 is squared off at this point. Photo by Kenneth M. Ardinger, collection of J. David Ingles.

M636

CP Rail 4717 illustrates an M636 for comparison of the radiator area. It rides on the MLW truck which resembles the GE floating-bolster truck. Photo by Larry Russell.

M630

British Columbia 726 illustrates the Comfort Cab or Safety Cab version of the M630, unique to that railroad. Photo by Larry Russell.

The M640, the only 18-cylinder locomotive in North America, was a one-of-a-kind experiment. It acquired a second life (including a new 18-cylinder engine) in February 1985 when it became a test bed for AC research. The standard MLW 3-motor truck was changed to an A1A wheel arrangement by fitting a pair of 1000-h.p. Brown Boveri FRA4073 three-phase asynchronous squirrel-cage motors. They are fed alternating current through four Brown Boveri thyristor convertors which occupy the space where the central air system had been, just behind the cab. Combustion air is drawn directly through carbody filters at the rear of the hood, and a smaller air system cools the large electrical compartment. The experiment is to continue through 1988.

A similar conversion is to be made at Beech Grove, Indiana, of Amtrak F40PH 202. Metro North plans to convert 10 FL9s to AC traction; New Jersey Transit is expected to re-equip 230 Arrow III multiple-unit cars with AC motors; and SEPTA (Southeastern Pennsylvania Transportation Authority) plans to order 26 cars with AC motors for the Norristown high-speed line. The New York City Transit Authority is also experimenting with the concept on two subway cars.

Why all this interest in AC motors? AC induction motors lack the brushes and commutator of DC motors — both high-maintenance items. Induction motors had previously meant a constant-speed locomotive (the older electric locomotives of the Virginian Railway were this type) but with modern electronics it is possible to create a current supply in which frequency and amplitude of the alternating current can be varied independently to produce uniform torque at varying speeds.

One order of HR616s was delivered: Canadian National 2100-2119. They have full-width bodies with Draper Taper. An order for 15 more, which were to have been 2120-2134, was canceled. With this cancellation, MLW announced it would quit the freight locomotive business but continue producing the LRC. CN 2104 photo by Wendell Lemon; CN 2114 photo by Gordon Lloyd Jr.

HR616

CP Rail 4744 has become a test bed for experimentation with AC traction motors. Photo by Gordon Lloyd Jr. **M640**

OTHER BUILDERS

FAUR SWITCHER: 1250 H.P., B-B

Model	AC/DC	H.P.	Cyls.	Length	Period produced	Approximate number of units sold		
						U. S. A.	Canada	Mexico
Quarter Horse	Hydraulic	1250	6	43'8"	1974	1	None	None

The Quarter Horse is a Romanian product marketed in the United States by Stanray Corp. The initials FAUR stand for Romanian words meaning, approximately, "Consolidated Rolling Stock Factory," a nationalized enterprise. The engine of the Quarter Horse is said to be a Sulzer design built under license, and the hydraulic transmission was a Voith product. The Quarter Horse demonstrated on several Class 1 railroads but gathered no orders. Because Morrison-Knudsen promoted the Sulzer engine in the U. S. and the Voith transmission was used on Amtrak turbos, the Quarter Horse was not quite the orphan it appeared to be — but in the end it was.

The demonstrator was sold in 1978 to the Washington Terminal Company. Later it was acquired by General Electric Railcar services and stored at its shop, the former Lehigh Valley Railroad shop at Sayre, Pennsylvania. It was still there in October 1986. Photo by Gordon B. Mott.

LIGHTWEIGHT-TRAIN POWER CARS

Builder	Transmission	H.P.	Cyls.	Length	Truck centers	Period produced	Approximate number of units sold U. S. A.	Canada	Mexico
United Aircraft	Hydraulic	1000	Turbine	73'9"	59'3"	11/67-11/68	4[1]	10[1]	None
ANF-Frangeco	Hydraulic	1140	Turbine	86'1"	55'8"	7/73-2/75	12	None	None
Rohr	Hydraulic[2]	1140	Turbine	87'2.23"	55'8.5"	7/76-11/76	14	None	None
MLW	AC	2900	12	67'11"	41'0"	8/73	None	1	None
MLW	AC	3725	16	65'2"	39'2"	2/80-9/84	2	31	None
Werkspoor	DC	2000	(2)16	79'6"	52'1"	5/77-8/77[3]	None	4	None

[1]Two of the Canadian units were subsequently resold to Amtrak.
[2]Westinghouse traction motor connected to hydraulic transmission for third-rail operation into Grand Central Terminal, New York.
[3]Secondhand delivery dates to Ontario Northland Railway after *Trans Europe Express* service; built 1957.

About every 20 years, after the retirement of the railroaders who remember why articulated lightweight trains are undesirable, along comes another generation that is captivated by the engineering advantages of greatly reduced train weight. Lightweight trains can achieve high speed on relatively little horsepower, and often have low centers of gravity and special suspensions for higher-than-normal speeds on curves. Such trains have been successful in Europe, where trains are frequent, but in the U. S., where trains are few, the expected passenger load fluctuates enough to make the fixed consists of lightweight trains awkward. Lightweight trains also tend to be expensive to maintain.

Amtrak, Canadian National, and CN's passenger successor VIA Rail Canada gave the idea another turn, however. Although some of the equipment mentioned has been retired, all is listed to preserve the record of this most recent round of lightweight experimentation.

The first of the 1967-1984 lightweight train attempts was the United Aircraft TurboTrain of 1967. Two 3-car trains were built by Pullman-Standard for United Aircraft for sale to the U. S. Department of Transportation. The power cars, two per train, were numbered 50-53. After tests, they began service between New York and

Boston on Penn Central's ex-New Haven line in April 1969. The trains continued in that service after the creation of Amtrak. They were retired in September 1975 because they were mechanically unreliable and had an availability of only 60 percent. They were scrapped in 1980.

At the same time the DOT sets were built, five 7-car trainsets were assembled in Canada by MLW for United Aircraft for delivery to Canadian National. CN intended to use them to replace conventional *Rapido* trains between Montreal and Toronto, but mechanical problems also vexed these sets. Not until 1973 was CN able to establish a reliable service using three of the original five pairs of power units spliced by seven intermediate cars. The three trainsets were conveyed to VIA Rail Canada. The power units carried CN numbers 125, 126, 129 (club seating), 150, 151, and 154 (coach seats). VIA renumbered the first three 145, 146, and 149. All were retired by 1983.

The other two pairs of power units and four intermediate cars were sold to Amtrak in 1973. One pair (which would have been numbered 54 and 55) was destroyed in a collision before delivery. The surviving set (56 and 57) was retired in 1975 along wth the Pullman-built sets.

Despite the difficulties experienced with the UA turbos, Amtrak was determined to experiment with two leased turbine train sets produced by ANF-Frangeco of France (there were 4 power units numbered 60-63). When these "Turboliners" proved moderately successful in service out of Chicago, 4 more train sets were ordered (the power units were numbered 58, 59, and 64-69).

At the same time, to avoid criticism for making non-American purchases, Amtrak arranged to purchase additional Turboliners to be built under license by Rohr Industries: 7 sets with 14 power units numbered 150-163. These sets were based at Rensselaer, New York, and are used in the New York-Niagara Falls corridor. Rohr went out of the rail car business after delivery of these trains. The Rohr and Frangeco trains experienced high operating costs compared to diesel-powered Amfleet trains, so it is unlikely Amtrak will experiment further with turbines.

In the meantime, MLW introduced a turbine-competitive diesel model, the M429LRC (for "Light, Rapid, Comfortable"). The locomotive and MLW-built cars had leveling devices to permit higher curve speeds. The demonstrator toured the U. S. and Canada, making several tests. Amtrak tried a pair of LRC diesels and 10 coaches in 1980 but found the curve tracking disappointing and returned them to the builder. VIA Rail Canada bought 31 LRC power units between 1981 and 1984.

Ontario Northland bypassed the home-built product in favor of secondhand *Trans Europe Express* equipment in 1977.

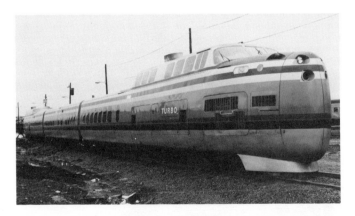 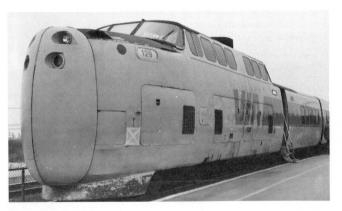

These photos offer a comparison of the Canadian- and U. S.-assembled UA TurboTrain power cars. The U. S. version hauled only 1 trailer between the power units; the Canadian trains originally had 5 trailers between the power units — 2 more were added in 1972. The extra cars came from the sets sold to Amtrak, so that Amtrak's ex-CN train had only 2 trailers between the power units. Wheel arrangement of the power cars is B-1. USDOT TurboTrain photo by Jack Armstrong; VIA TurboTrain photo by Gordon B. Mott.

UA Turbotrain

ANF-Frangeco Turboliner

The French-built turboliner can be easily distinguished from the U. S.-built version by its three-piece windshield and protruding headlights. The wheel arrangement of the power car is B-2. As of November 1988 three of the later group of Frangeco sets had just been rebuilt for service between New York, Albany, and Niagara Falls. The rebuilding includes new noses similar to those of the Rohr Turboliners. TRAINS Magazine photo by J. David Ingles.

Rohr Turboliner

The Rohr Turboliner has a two-piece windshield and a more streamlined appearance than the French-built train. The wheel arrangement of the power car is B-2. As of November 1988 all of the Rohr sets were still in service between Niagara Falls, Buffalo, Albany-Rensselaer and New York's Grand Central Terminal. Photo courtesy of Amtrak.

LRC

The original LRC demonstrator and a prototype tilting coach intended to allow higher speeds by leaning outward on curves were produced in 1973 and demonstrated in the U. S. and Canada. Amtrak thought this might be a cheaper solution to the curvy ex-New Haven route than electrification, and acquired 2 units, 38 and 39, built in February and June 1980, and 10 coaches. However, the tilting car feature was not as successful as hoped, and the equipment was returned to the builder. VIA LRC diesels 6900-6920 were built from February 1981 through June 1982, and 6921-6930 from June 1983 through September 1984. The 1973 demonstrator had small windshields and squared openings for the headlights and ditch lights. The builder retained the demonstrator and the two ex-Amtrak units. Photo by Gordon B. Mott.

LRC

Amtrak 38 and 39 had the larger windshields and unfaired headlights of the subsequent VIA engines, but lacked the ditch lights of the Canadian units. Photo by Herbert H. Harwood.

Number 6917 is a typical VIA LRC. Curiously, the LRC design does not include a shroud or housing for the coupler, unlike most lightweight train designs of the past. Photo by Gordon Lloyd Jr.

LRC

The secondhand Trans Europe Express (TEE) sets which the Ontario Northland placed in service in 1977 were four of five built in 1957 (the fifth set had been scrapped in 1971 after a derailment). Each set consisted of a diesel-electric power car with A1A-A1A wheel arrangement, and three passenger cars, the third of which had a control cab for push-pull operation. The cars were connected with standard European screw couplers, and full-width diaphragms filled the space between cars. The total length of the train was 321'8.8". The sets were designed to be run in multiple. The trailing cars were built by Swiss Industrial Company of Neuhausen Rhine Falls (SIG), the electrical equipment by the Swiss firm Brown Boveri of Baden, and the power cars by Werkspoor of Amsterdam, Netherlands.

Two sets — 501 and 502 — were owned by Swiss Federal Railways (501 was scrapped in 1971) and three — 1001, 1002, and 1003 — by Netherlands Railways. The longest-lasting assignment for the trainsets was between Amsterdam and Zurich. Two sets were required, and they remained on that run until 1973. Other sets operated Amsterdam-Brussels-Paris and later Zurich-Paris and Zurich-Munich. The trains disappeared from the TEE listings in *Cooks Continental Timetable* in 1973.

The four surviving sets were sold to Ontario Northland in 1977. ONR chose not to use the cab cars to lead because of concern for derailment in the event of a grade crossing collision, but it occasionally used the sets in multiple. On arrival, ONR numbered the first two sets 1900 and 1901 but soon renumbered them 1980 and 1981. The second two sets were numbered 1982 and 1983. These numbers were carried in the indicators on both the power car and the cab car and are the number of the set, not the individual car.

The power cars had two 16-cylinder engines rated at 986 gross horsepower each; approximately 1775 h.p. was available for traction by North American methods of calculation. Predictably, the 20-year-old Werkspoor diesels did not last long in service, given North American maintenance practices. Failures were frequent.

ONR converted FP7 1519 to power car 1984 and placed it in service in December 1979 in place of one of the Werkspoor units. The 300 h.p. Werkspoor auxiliary diesel that powered the HEP alternator was replaced by a 300 h.p. Harper-Detroit diesel placed in the former steam-generator compartment of the FP7. European buffers and coupler were mounted on the rear of the FP7. Subsequently FP7s 1518, 1501, and 1510 became 1985, 1986, and 1987, and the Werkspoor power cars were scrapped in 1983.

The passenger cars were in superb condition and have been kept that way by ONR, which continues to use them daily in *Northlander* service from Toronto to North Bay on Canadian National and from North Bay to Timmins on ONR, an 11-hour trip in all. The control cab cars remain as reminders of the appearance of the Dutch locomotives, even though the cabs always trail and are never used for control. As only two sets are required to protect normal service, there is ample time for maintenance and the sets are alternated in and out of service weekly.

On occasion Ontario Northland coupled trainsets together and ran them in multiple. In this photo the power cars are at the outer ends of the train. There is no passageway between the two control cab cars in the middle of the train. TRAINS Magazine photo by J. David Ingles.

ONR FP7 1510 became power car 1987. The number is also carried in the indicator of the trailing cab control car. Photo by Greg Sommers.

CAPITAL REBUILD PROGRAMS

A Capital Rebuild Program (CRP) is an accounting device that allows railroads to treat rebuilt locomotives as if they were new investments. To qualify as a CRP under Interstate Commerce Commission and Internal Revenue Service rules, the rebuilding costs must exceed half the original purchase price of the locomotive. The railroad can then depreciate the rebuilt locomotive over its entire anticipated additional life, usually 8, 10, or 15 years. Otherwise, the rebuilding expenses must be treated as maintenance costs in the year the repairs are made and there are no tax advantages.

Until recently a CRP qualified for investment tax credit. The recent elimination of that point of the tax law will affect future programs, but rapid changes in other aspects of the locomotive market make it difficult to predict the results. However, because most railroads used up their allowable investment credits on other projects, they often sold the rebuilt locomotives to a lessor who then leased the locomotives back to the railroad at a lower cost than the railroad would have to pay to borrow cash for repairs. Such leases are profitable to both parties because the lessor receives an investment tax credit and shares the benefit with the railroad through a lower interest rate on the lease. These leverage leases theoretically allow the railroad to take title to the locomotive at the end of the lease, but in most cases the railroad surrenders the locomotive to the lessor (lease turnback) because of the high buyout price or because major overhaul expenses often coincide with the expiration of the lease. The decision to buy out at the end of these leases or to turn back the locomotives depends on the used-locomotive market and the price of new locomotives.

The best candidates for rebuilding are EMD switchers and hood units — GP7, GP9, SD7, or SD9. Few GE locomotives are old enough to qualify for rebuilding, and other non-EMD makes, being out of production, are poor long-term prospects for parts availability, even if their maintenance cost history justifies rebuilding. On a typical large railroad there are many low-mileage jobs for which a GP/SD/7/9 is ideal. Capital rebuilding offers an intermediate economic choice between continued maintenance overhauls of elderly GPs and SDs and replacing them with new units. EMD's GP15-1 is a builder's response to the competition of capital rebuilding programs.

The standard package which characterizes capital rebuildings of EMD units usually includes upgrading to "E" power assemblies, but this is almost prohibitively expensive with "B" engines. If a GP7 or SD7 is the candidate, the railroad may swap around to provide a C engine on which the E heads will fit, or settle for a less-satisfactory "BC" conversion which retains the old B head and uses a water jumper to solve the B engine water leak problems. A minor increase in horsepower often results, especially if the package includes the four-stack "liberated exhaust" system pioneered by Missouri Pacific.

Rebuilding may include complete rewiring to eliminate deteriorated insulation. In most cases the electrical system is rebuilt using large old-style contactors instead of the more recent modular apparatus. Most CRP locomotives are given central engine air filtration systems, using a prominent hood-top air intake and filter assembly.

Generally, the air brakes on a CRP engine are upgraded to the 26L type, although sometimes a minimal 6BLC-type conversion is made, which allows operation with units equipped with 26L or 24RL brakes. Peripheral changes such as chopped noses, retention toilets, and FRA-mandated safety glazing, footboard removal, and related pin-lifter modifications are done. Radiators, air compressor, and fuel and water pumps are reconditioned, as would be done in any heavy overhaul.

The future of CRP programs depends on several things besides tax law changes. Most important is the question of repair parts. EMD naturally gives first preference to new locomotives in its parts production. There is a fixed pool of older parts that can be reconditioned, but some of those are getting beyond the point of economical reconditioning. Some reconditioned parts furnished by other vendors have been of low quality. There are also questions of whether the railroad's shop forces can perform the work competitively, and whether the cost is competitive with a simple heavy overhaul.

EMD SWITCHERS REBUILT BY RAILROADS

Railroad	Old model	New model	Engine	Cyls.	H.P.	Period produced	Approximate number of units produced	Shop location
IC	SW1	SW1R	645E	8	1000	10/68	1	Paducah, Kentucky
IC/ICG	TR2, NW2	SW13	567BC	12	1200	1971-1975	12 cab, 3 booster	Paducah
ICG	NW2, SW7, SW9	SW14	567BC	12	1200	3/78-3/82	112	Paducah
BN	NW2	NW12	645E	12	1200	2/75-10/76	4	W. Burlington, Iowa
UP	SW7, SW9	SW10	567BC	12	1200	9/79-12/84	75	Omaha, Nebraska
AT&SF	NW2, SW9	SSB1200	567BC	12	1200	1/74-12/79	29	San Bernardino
CNR	SW1200, GP9	GS-413	645E	12	1200	1985-1987	18	Pt. St. Charles, Que.

SW1R

The 645 repowering of Illinois Central SW1 612 to SW1R 13 in 1968 was marked by a larger hood and two closely spaced stacks. The project was not repeated, but the prototype was successful and remained on the roster in 1986. Photo by Paul C. Hunnell.

SW13

SW13

IC moved on to the SW13 program which included the creation of a sealed hood with a central air filter system, even if it did not involve a new engine. These, too, were all on the roster in 1986. Note the vertical central air filter just ahead of the cab on 1301 and in a similar position on calf unit 1300B. IC 1301 photo by Greg Sommers; IC 1300B photo by Mark Carron.

All SW14

The SW13 program was followed by a simpler rebuilding program that resulted in the mass-produced SW14. The rebuilding did not include a sealed carbody, but did include a new cab with angular roof. The SW14s are dispersed throughout the Illinois Central system. Photo by J. David Ingles.

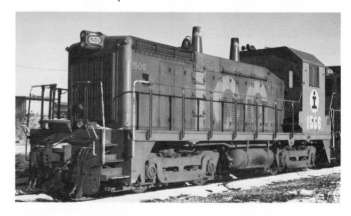

From engine 1421 on, the radiator treatment was modified to include exposing the ends of the radiator core assemblies at the top front of the hood, as on the SW1000 and SW1500. ICG 1506 (photo by Louis A. Marre) and 1491 (photo by Greg Sommers) illustrate the left and right sides of the later SW14.

One SW14 was produced for an outside buyer. East Camden & Highland No. 60 was built in 1981. Photo by Louis A. Marre.

Burlington Northern began an ambitious program to rework old NW2s, but completed only 4 before deciding that the program was not economical. A front number board and headlight enclosure similar to that found on the SW1500s identifies these units, as well as their low road numbers 1, 5, 14, and 19. Photo by Louis A. Marre.

The first units out of Union Pacific's SW10 program were numbered in the 1800 series, but they were soon re-numbered to the 1200s (1200-1274). The salient mechanical feature of the rebuilding is the substitution of salvaged GP7 and GP9 radiators and electric fans for the original mechanically driven front fans. Photo by George Cockle.

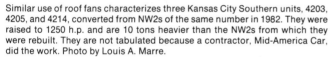

Similar use of roof fans characterizes three Kansas City Southern units, 4203, 4205, and 4214, converted from NW2s of the same number in 1982. They were raised to 1250 h.p. and are 10 tons heavier than the NW2s from which they were rebuilt. They are not tabulated because a contractor, Mid-America Car, did the work. Photo by Louis A. Marre.

Amtrak has used roof radiator fans on some of its switcher rebuildings. One example is ex-Conrail, ex-Lehigh Valley SW8 No. 750, which was originally one of LV's dynamic-brake-equipped SW8s, accounting for the evidence of sheet metal work being done at the rear of the hood. Amtrak's heavily re-worked SW1s share this radiator fan feature. They are not tabulated because Amtrak does not qualify for investment tax credit, and thus this is not a true capital rebuild program. Photo by Greg Sommers.

SSB1200

anta Fe's SSB1200 switcher rebuild program (the model designation stands for Switcher, San Bernardino) qualifies for mention here because of the uprating to 1200 h.p. of 3 NW2s as part of the program. Other units rebuilt or upgraded between 1974 and 1979 were 23 SW9s, 3 SW1200s, and 2 TR4 sets. All were retired in 1984 and 1985; most were sold to other railroads. Photo by Vic Reyna.

One of the more unusual switcher rebuild programs was undertaken by Canadian National at Point St. Charles Shops in Montreal in 1985 and 1987. The design included SW1200RS frames and GP9 hoods. The result, CN class GS-413, is rated at 1300 h.p. and is not likely to be confused with anything else. Since CN continues to carry out heavy overhauls on its SW1200RS fleet, the purpose and future of the program is not clear. These 18 units are tabulated because of the significant alteration in appearance, though this is not a true capital rebuild program because of the difference in Canada's tax laws and because Canadian National Railways is government-owned. Photo by Pierre Patenaude.

GE 70-TONNER REBUILT BY RAILROAD

Louisville & Nashville GE 70-tonner 98 was repowered in 1966 with an 800 h.p. Alco 6-251B engine. Somewhat surprisingly, when the locomotive was severely damaged by fire, it was rebuilt again by South Louisville shops in June 1978. It was then sold in early 1980 to Tropicana to switch its Kearney, New Jersey, warehouse. A second 70-tonner, No. 99, was similarly repowered and disposed of earlier. Photo by Tom Trencansky.

EMD ROAD-SWITCHERS, REBUILT BY RAILROADS, 1500-2000 H.P., B-B

Railroad	Old model	New model[1]	Engine	H.P.	Cyls.	Length	Truck centers	Period produced	Approximate number of units produced	Shop location
MP	GP7	GP7	567C, BC	1800	16	55′11″[2]	31′0″	ca. 1968	37	North Little Rock, Ark.
MP	GP18	GP18	645E	1900	16	56′2″	31′0″	ca. 1969	1	North Little Rock, Ark.
MP	GP18	GP18	645E	2000	16	56′2″	31′0″	ca. 1969	5	North Little Rock, Ark.
MILW	GP9	GP20	645E	2000	16	56′2″	31′0″	6/69-2/70	18	Milwaukee, Wis.
MILW	GP9	GP20	645E	2000	16	56′2″	31′0″	4/72-10/73	36	Milwaukee, Wis.
IC/ICG	GP7	GP8	567BC	1600	16	55′11″[3]	31′0″	9/67-1977	109[4]	Paducah, Ky.
IC/ICG	GP9	GP10	567C	1850	16	56′2″[3]	31′0″	6/67-1977	327[4]	Paducah, Ky.
ICG		GP11	567C	1850	16	56′2″	31′0″	4/78-1981	54[4]	Paducah, Ky.
BN	GP7	GP10	645E	1800	16	55′11″	31′0″	7/74-11/76	26	West Burlington, Iowa
AT&SF	F7	CF7	567BC	1500	16	55′11″	31′0″	2/70-3/78	233	Cleburne, Texas
CN	GP9	heavy GP9	645E	1850	16	56′2″	31′0″	9/81-7/84	33	Pt. St. Charles, Que.
CN	GP9	light GP9	645E	1850	16	56′2″	31′0″	4/84-9/84	9	Pt. St. Charles, Que.
CN	GP9	yard GP9	645E	1850	16	56′2″	31′0″	6/85-6/86	17	Pt. St. Charles, Que.
C&NW	GP7	HE15	KTA 3067	1500	16	55′11″	31′0″	4/80	2	Oelwein, Iowa

[1]These are unofficial model designations by the railroad, except where the original designation has been retained.
[2]MP GP7s 309-331 and 333-336 were built in 1954 after the completion of regular GP7 production and had GP9 frames.
[3]Due to swapping of frames, some GP10s may have 55′11″ GP7 frames and some GP8s may have 56′2″ GP9 or GP18 frames.
[4]Totals do not include units built for customers other than IC and ICG.

Missouri Pacific 1726 is a typical upgraded MP GP7 of the 1960s, equipped with a four-stack "liberated exhaust" manifold and rerated to 1600 h.p. from the original 1500. Such modifications were typically done in the course of regular maintenance at North Little Rock shops, and are therefore not tabulated among the capital rebuilds. By 1968, GP7s 68-298 had been converted to 1600 h.p., as had F3s 743-773 and F7s and FP7s 785-942; and GP9s 340-399 had been raised from 1750 to 1800 h.p. The GP7s were renumbered in 1975 to series 1600-1749, including No. 1726 illustrated, ex-269. Photo by James B. Holder.

MP passenger GP7s 300-336 apparently were also upgraded to 1600 h.p. In 1968, as shown in the table of capital rebuilds, they were selected for further uprating to 1800 h.p. They were probably chosen for this conversion because with rooftop air reservoirs they had room for a 2500-gallon fuel tank between the trucks. Some units such as No. 302 also received low noses in rebuilding. This group was renumbered 1750-1786 in 1975; the units have subsequently been traded in or scrapped. Photo by Lee Hastman.

Other former MP passenger GP7s retained their high short hoods: 314, 320-325, 327-330, 333, and 336. MP 1765, ex-314, illustrates this, and helps make the point that significant rebuilding may occur without the "capital rebuild look," while in turn, many a low-nose GP9 "rebuild" is merely a classified repair accompanied by a low nose. Photo by James B. Holder.

The next step in MP's rebuilding activity was to use rooftop air reservoirs to increase fuel capacity of GP18s 505-530, built in 1960. However, MP installed 2000-gallon tanks instead of the 2500-gallon tanks used on the ex-passenger GP7s. At the same time, North Little Rock lowered the noses of 505-523, 525, 526, 528, and 530. All except 530 kept their 567D1 power assemblies and 1800 h.p. rating, but 530 was experimentally fitted with 645E assemblies and given a 1900 h.p. rating. This unit was renumbered 1878 in 1975 and sold to Precision National in June 1985. Photo by J. David Ingles.

MP GP18s 531-550 were delivered in 1962 with low noses and 2000-gallon tanks. MP put 645E assemblies in the first five of these, 531-535, and raised their ratings to 2000 h.p. They were numbered 1879-1883 in 1975 and sold to Precision in July 1985. Photo by James B. Holder.

Milwaukee Road 999 was the first of that railroad's capital rebuild program. Externally, the chopped nose and a GP20 model plate were the only evidence of the change. The series was numbered downward from 999. MILW 971 illustrates the second group, which incorporated the central engine air-filter package with weather hoods, and also employed a larger number box above the windshield. Both photos by Louis A. Marre.

Illinois Central's Capital Rebuild Program, carried out at the road's Paducah, Kentucky, shops, was a pioneer — the first such program and an extensive one. GP7s were rebuilt and designated GP8s; GP9s became GP10s. The first examples of each type were GP8 No. 7960 and GP10 No. 8109. They and GP10s 8009, 8025, and 8082 retained their high noses. The first chopped-nose Paducah rebuild was 7961, outshopped in March 1968, and the Paducah rebuilds had this feature thereafter.

During 1968 there was a brief flirtation with rating the rebuilds at 2000 h.p. Those so rated — but only temporarily — were GP8s 7957, 7964, 7966, 7971, and 7981 and GP10s 8004, 8025, 8072, 8082, 8158, and 8233. They were all derated to 1850 h.p. later. The five 7900s are included in the table as GP8s; considering their horsepower rating, they should be GP10s. Exchanges of C and B engines occur frequently in the course of normal maintenance, so other such crossovers are likely to take place. Normal 1850 h.p. production of the GP10 resumed in March 1969 with No. 8031, and production of the normal 1600 h.p. GP8 resumed with 7968 in April.

Number 7968 was also equipped with paper air filters, indicated by the hood-top box, as were subsequent GP8s and GP10s. All the uprated units have the four-stack liberated exhaust manifold.

ICG removed dynamic braking and eliminated the characteristic blister on units rebuilt for its own use. None of the IC's own GP7s or GP9s was originally so equipped, but many secondhand units were.

IC was pleased with the performance of its rebuilds and decided Paducah Shops should go into the business of upgrading for others as well. IC set up a joint venture in which IC did rebuilding for Precision National Corporation (PNC) of Mount Vernon, Illinois.

At the same time IC began to work through Precision to obtain secondhand GP7s and GP9s to rebuild and add to its own roster. Of the GP8s in the table, 59 are based on secondhand units acquired by IC and ICG. Of the GP10s, 60 are secondhand. It is difficult to attribute rebuilt units to particular original locomotives, since there was free mixing of components when a group of units was on the floor. The official identification of the predecessor of a given rebuild was a matter of accounting. For example, ICG 8270, a GP10 outshopped in December 1971, was built on the frame of wrecked Boston & Maine GP9 1710 purchased from Precision, but incorporated the engine and components from wrecked Southern Pacific GP9 No. 3516. SP 3516 was made the official predecessor, while B&M 1710 was charged against 8016, outshopped in March 1977.

IC's merger of Gulf, Mobile & Ohio had no effect on the rebuilding program because GM&O had no GP7s or GP9s. ICG considered using GM&O's shop at Iselin, Tennessee, as a second CRP facility because of conflicts between contract work and ICG's own program. ICG, however, decided to do the work solely at Paducah, so there were long gaps in the ICG's program because contract jobs preempted Paducah's floor space.

The numbering system used with the Paducah rebuilds is hard to summarize, because it changed in midstream. Originally the numbers of all rebuilds were 1000 lower than the original IC numbers; for example, 7960 was the former 8960. The 7700 series was begun for foreign-road GP7s being rebuilt into GP8s, while series 8267-8299, 8390-8399, and 8429-8499 were opened for this purpose for GP10s. Numbers 7982-7999 were also set aside for GP8s.

For the most part numbering followed this pattern until 1977, when ICG began to assign vacant numbers regardless of former number or owner, apparently in an effort to make solid blocks of numbers that would be easier to deal with in documents. Units completed during the program were:

GP8: 7700-7746, 7800, 7850-7852, 7900-7918, 7950, 7952-7955, 7957, 7960-7972, 7974, 7976, 7978-7979, 7981, 7983-7999 (7900 was renumbered from 7982; 7957, 7964, 7966, and 7971 were later rerated to GP10 specifications)

GP10: 8000-8171, 8173-8181, 8183, 8184, 8186-8197, 8199, 8202-8205, 8207, 8209-8217, 8219-8230, 8233-8237, 8240, 8241, 8243, 8245, 8247-8251, 8254, 8256-8260, 8265-8283, 8285-8296, 8300-8332, 8335, 8337, 8339, 8343, 8346-8348, 8351, 8354, 8357-8359, 8361, 8362, 8365, 8370, 8371, 8375, 8377, 8379, 8383, 8387, 8390-8397, 8418, 8442-8447, 8460-8466

Other railroads whose GP7s, GP9s, and GP18s became the basis of ICG GP8s and GP10s include Detroit, Toledo & Ironton; Pittsburgh

& Lake Erie; Reading; Frisco; Quebec North Shore & Labrador; Chesapeake & Ohio; Baltimore & Ohio; Boston & Maine; Denver & Rio Grande Western; Clinchfield; Florida East Coast; and Union Pacific. The influx of foreign units provided ICG with sufficient power to handle traffic growth and retire all Alco diesels inherited from Gulf, Mobile & Ohio. ICG's Capital Rebuild Program was terminated in 1981, and the Paducah shop was later sold to the Paducah & Louisville Railway, whose subsidiary, VMV Enterprises, operates it. Diesel locomotive rebuilding has resumed there.

The first GP8 from IC's rebuilding program was No. 7960. Photo by James B. Holder; collection of Kenneth M. Ardinger.

GP8

Illinois Central 8109 was the first GP10 produced by the railroad's Capital Rebuild Program. Photo by J. David Ingles.

GP10

GP8

The first Paducah rebuild with a chopped nose was GP8 7961, outshopped in March 1968. Only five Paducah rebuilds retained their original high short hood. Photo by W. S. Kuba.

GP8

GP8 No. 7968 was equipped with paper air filters, attested to by the box atop the hood above the letters "IL." Photo by J. W. Stubblefield; collection of J. David Ingles.

GP10

There is little difference between the 567 C and D engines, so a GP10 can be based on a GP18 as well as a GP9. ICG 8418, rebuilt from a wrecked GTW GP18 and outshopped in July 1974, illustrates this. Photo by Alton B. Lanier; collection of J. David Ingles.

On its own GP8s and GP10s, ICG originally favored a Pyle Gyralite oscillating headlight mounted over the windshield. Later, to simplify maintenance, ICG shifted to a solid-state warning light as shown on 8708. The white lights in the top and middle positions alternate. One is aimed down and to the right and the other down and to the left, simulating the sweep of the Gyralite. The third light is red, and comes on automatically in an emergency brake application, signaling trains which might be passing on double track to stop. The last unit with a Gyralite was 8307, and the first with a solid-state light was 8444, both outshopped in January 1974. Photo from Illinois Central Gulf Railroad.

The final group of Paducah rebuilds had angled cab roofs, illustrated by No. 8314. The units were based on Union Pacific GP9Bs, and Paducah had to build new cabs from scratch. The first of those units was 8462, produced in June 1977. GP8 7800 also has such a cab, although it was nominally based on a Pittsburgh & Lake Erie GP7. TRAINS Magazine photo by J. David Ingles.

Starting in March 1977, ICG rebuilds incorporated a more compact form of central air filter, the single-stage Dynacell, as shown on ICG 8016, the first unit so equipped. Photo by W. S. Kuba.

GP11

ICG 8301 is the prototype GP11, built in 1978. Using Dash 2 electrical equipment required complete reworking of the cab. Installing a central equipment blower required relocating the cab one foot forward. The equipment blower duct is on the left side behind the cab; the duct to the rear traction motors is on top of the left-side running board. Photo from Illinois Central Gulf Railroad.

GP11

Subsequent production of GP11s consisted of 8701-8726 in 1979, 8727-8750 in 1980, and 8751-8753 in 1981. Note that the blower duct bulge is vented on the production units such as 8706, compared to prototype 8301. Photo by Greg Sommers.

GP10

Burlington Northern's GP10s were "Cadillacs," so much so that resumption of the program could not be justified after the 1975 recession. All capital went toward new SD40-2s and C30-7s for coal service. Two partially rebuilt units were released in 1976, and BN went back to overhauling GP7s in kind. Hooded paper air-filter boxes and four exhaust stacks make BN's 1400s easy to spot. Photo by Kenneth M. Ardinger.

One of the most unusual rebuildings — nearly a new locomotive — is Santa Fe 1160, rebuilt in December 1970 from Baldwin VO 1000 No. 2220, using a GP7 hood, a 16-cylinder 567B engine rated at 1500 h.p., an EMD generator, and EMD Blomberg trucks. The locomotive was produced by AT&SF's Cleburne, Texas, shops and can be considered a dress rehearsal for the CF7 program. The rebuild was originally numbered 2450; later than the date of the photo it was again renumbered 1460 and eventually served as shop switcher at Cleburne. Canadian National's similar-looking GSW-13s have 12-cylinder engines. Photo by James B. Holder.

GP7U

Santa Fe had an extensive program of overhauling its GP7s and GP9s in kind at Cleburne, Texas. The GP7 program ran from November 1972 through December 1981 (242 units), and the GP9 program ran from January 1978 through May 1980 (56 units). The latter included creating GP9 2244 from GP7B 2798A. Number 2064 illustrates the initial phase of the program, with four-stack manifold, 567BC engine and lowered nose, but with the old cab retained. Units 2000-2004, 2050-2067, 1310, and 1311 look like this (the 1300s are slug control units). In 1974, ATSF adopted the angular "Topeka cab," as shown on No. 2216, to permit the installation of air conditioning. GP7s 2005-2027, 2068-2243, and 1312-1329, and GP9s 2244-2299 are so equipped. Dynamic braking was removed from units that had it and the blister was blanked off. Both the GP7 and GP9 programs were carried out at Cleburne, Tex. Both photos by Greg Sommers.

The Santa Fe CF7 (converted F7) is unique among CRP locomotives. Because of the bridge-truss design of the F7 body, the railroad had to manufacture new underframes. In the original configuration, illustrated by No. 2647, the reconstructed cab retained the old F7 roof line and used the F-unit side windows. The first CF7, 2649, had dynamic braking because the salvaged GP7, used for its hood, had it. All later CF7s, numbered down from 2649, had homemade hoods like that on 2647 and lacked dynamic brakes. Photo by Charles M. Mizell Jr.

CF7

CF7

AT&SF 2522 illustrates the intermediate phase of CF7 production, with four exhaust stacks, partially boxed-in underframe, and new side windows. The first unit with four stacks was No. 2612, rebuilt in 1972. In 1974, starting with No. 2470, a flat cab roof was substituted to accommodate air conditioning — there were 179 round-roof CF7s and 54 flat-roof versions. Number 2452 was fitted with 645 power assemblies, but this experiment did not persuade AT&SF to abandon 567 dimensions for the CF7s. CF7s 2612-2625 were fitted as RCE locomotives for potash and sulfur trains out of Clovis, New Mexico. Santa Fe's CF7s were all retired by 1988, many of them sold for service elsewhere. Both photos by J. R. Quinn.

GR-418

GY-418

The CN "heavy," "light," and "yard" GP9 rebuilds look alike. The heavy units bear road numbers 4000-4032; the light units, 4100-4108; and the yard units, 7200-7216. In addition, the light units have smaller fuel tanks, and the yard units are in class GY-418 rather than the GR-418 of the road units. It is probable that the field-limit feature has been altered on the yard units to make them better at kicking cars, a limitation otherwise encountered in using GP9s in yard service. In all three groups, dynamic brakes were eliminated and air filters were installed in the former dynamic brake blister. CN 4024 photo by Wendell Lemon; CN 7205 photo by Larry Russell.

Chicago & North Western's Oelwein, Iowa, rebuild program was not primarily an upgrade program. Rebuilt units retained their original GP7 or GP9 designation and rating. The distinction between rebuilt units and those given light overhaul and repainted was mainly a matter of whether sheet metal work was included, mainly nose lowering. Dynamic brakes were removed, and some units received liberated exhaust manifolds. C&NW applied the "rebuilt" designation only to its own units (73 GP7s and 52 GP9s), and not to any of the foreign units acquired from Precision, which are probably leased and therefore not eligible for capital rebuild accounting.

The tabulation below by Bruno Berzins is based on Chicago & North Western records. All are C&NW units unless noted otherwise.

4100-4209: ex-Rock Island GP7
4210-4251: vacant or assigned to Alcos
4252-4253: Cummins-powered "HE15"
4254-4278: vacant or assigned to Alcos
4279-4299: rebuilt GP7
4300: vacant
4301-4309: rebuilt GP9
4310-4319: rebuilt GP7
4320-4326: rebuilt GP9
4327-4332: rebuilt GP7
4333: rebuilt GP9

4334-4338: rebuilt GP7
4339: vacant
4340-4358: ex-Quebec North Shore & Labrador GP7
4359-4378: ex-Frisco GP7
4379-4399: ex-Union Pacific GP7
4400-4424: GP15-1
4425-4430: vacant
4431-4465: GP7s acquired from other railroads (Conrail, Frisco, C&O, QNS&L, D&RGW, RI) through Precision National

4466-4495: rebuilt GP7
4496-4499: ex-Union Pacific GP7
4501-4504: rebuilt GP9
4505: rebuilt GP7
4506-4513: rebuilt GP9
4514-4528: ex-Quebec North Shore & Labrador GP9
4529-4549: rebuilt GP9
4550-4559: ex-Rock Island GP9
4560-4562: rebuilt GP9

The rebuilding program left such items as air reservoirs and fuel tanks as they were. Number 4292 has its original small fuel tank and air reservoirs below the frame. Photo by Roger Bee.

Chicago & North Western's rebuilding sometimes included liberated exhaust, occasionally with three stacks, as shown on No. 4484. Photos by George Cockle.

Unquestionably a capital rebuild was C&NW's creation of Cummins-engined "HE-15" locomotives from GP7s 1534 and 1596. Number 1534 emerged from Oelwein as No. 4200 in April 1980; 1596 was completed as 4201 in August. They were subsequently renumbered 4252 and 4253. After late 1983 they were stored. Problems reportedly arose not with the 16-cylinder Cummins prime mover but with the generator (perhaps an engine-to-generator mismatch) and the belt-driven accessories. The units were last reported held at Oelwein for possible reworking. Photo by Bruno Berzins.

Southern Pacific had an extensive program of overhauling GP9s, calling them GP9Es and changing their road numbers. The units received central air filters, but as there was no engine modification, they are not tabulated as rebuilt. Photo by Vic Reyna.

Conrail relied mainly on contract rebuilds of its fleet of GP7s, GP9s, and GP18s, but CR 7479 illustrates an Altoona-rebuilt unit that has new wiring, an overhauled engine, central air filter, and blanked dynamic brake blisters. Photo by Louis A. Marre.

211

Seaboard Coast Line's GP16 program, conducted in the ex-Atlantic Coast Line Uceta shop at Tampa, Florida, was a modest upgrading program, using the four-stack exhaust manifold to increase GP7 horsepower rating to 1600, and cutting down the short hoods. It was basically a maintenance overhaul and is not tabulated as a rebuilding program. Because Uceta's facilities were limited, remanufactured engine components were shipped in from Waycross, Georgia, and remanufactured electrical components came from GE at Chamblee, Ga. The 1600 h.p. rating was maintained whether the original unit was a GP7 or a GP9. The units were sent to Waycross for painting. The program began in May 1979 and was halted in November 1982 due to economic conditions. By that time, the output consisted of locomotives 4600-4645 (BC engines, no alignment blocks for the couplers) and 4700-4809 and 4975-4979 (C engines, alignment blocks). CSX has numbered these units 1706-1860. Photo by Warren Calloway.

Uceta GP16

Canadian Pacific began programs of heavy overhauls at Ogden Shop in Calgary and Angus Shop in Montreal in 1980. The programs, still going at press time, involve GP7s, GP9s, and RS-18s. They have the short hoods lowered and maintain their former horsepower rating. Number 1578 is a representative GP9, and No. 1800 was the first RS-18 completed. By mid-1988, 12 GP7s, 157 GP9s and 47 RS-18s had been completed, with 10 more RS-18s scheduled for later in the year. CP's roster still has 22 RS-18s and 45 GP9s as potential material for rebuilding. CP 1578 photo by Wendell Lemon; CP 1800 photo by Gordon Lloyd Jr.

DETURBOCHARGING

A feature of many capital programs is removal of turbochargers from EMD's early turbocharged models: GP20, GP30, GP35, SD24, and SD35. This is usually done to avoid the problems of turbocharger maintenance on locomotives that have been downgraded to low-mileage, low-speed service. The standard Roots blower replaces the turbocharger, and the engine is upgraded with 645 power assemblies. The resulting locomotive usually has a 2000 h.p. rating.

Penn Central pioneered deturbocharging with its conversion of ex-New York Central GP20s 2100-2112 in 1969 and 1970. At the same time, paper air filters were added — one of the earliest applications of this now-common aspect of conversion. The resulting unit looks like a GP20 except for the four-stack liberated exhaust manifold replacing the single turbocharger stack, and the projecting air filter box on top of the hood. The units, which kept the same numbers under Conrail ownership, have been retired.

Subsequent deturbochargings included Southern Pacific and Cotton Belt GP20s, GP30s, and GP35s, Missouri Pacific GP35s, and Illinois Central Gulf and Chicago & North Western conversions of ex-Union Pacific and ex-Southern SD24s.

Deturbocharged GP20

The clue that this GP20 has been deturbocharged is the paper air filter box in the place where the turbocharger stack would normally be. Photo by Vic Reyna.

Missouri Pacific's deturbocharging of GP35s began in 1974 with the rebuilding of GP35 630, which had been damaged in a wreck. This unit became 2007, then was renumbered 2008 shortly thereafter. In the 1975 renumbering, the GP35s took numbers in the 2500 series, and the following 16 were similarly deturbocharged: 2501, 2503, 2505, 2506, 2507, 2512, 2513, 2516, 2519, 2521, 2523, 2530, 2541, 2543, 2551, and 2560. All 17 were renumbered 2601-2617 in April 1978; No. 2008 became 2600. These units were retired in 1986, and many were sold for service elsewhere, for example, Kiamichi 3803, ex-MP 2607. The GP35 fans remain unchanged but four exhaust stacks and an air filter box betray the lack of a turbocharger. Photo by Louis A. Marre.

Apparently the start of a program that would parallel the deturbocharging of SD24s to create SD20s, Illinois Central Gulf removed the turbochargers and added 645 power assemblies to ex-Gulf, Mobile & Ohio GP35s 502 and 514 in March and April 1981, creating 2250 h.p. 2601 and 2602. ICG's Capital Rebuild Program was terminated in 1981, making the two units the only ones of their type on ICG. Alco trade-ins back in 1965 account for the Type B trucks. Photo by Greg Sommers.

EMD ROAD-SWITCHERS, REBUILT BY RAILROAD, 1500-2600 H.P., C-C

Railroad	Old model	New model[1]	Engine	H.P.	Cyls.	Length	Truck centers	Period produced	Approximate number of units produced	Shop location
UP	SD24		645E	3300	16	60'8"	35'0"	7/68	1	Omaha, Nebr.
UP	SD24	SD24-4	567D	2400	16	60'8"	35'0"	5/75	1	Omaha, Nebr.
MILW	SD7, SD9	SD10	645E	1800	16	60'8"[2]	35'0"	3/74-1/76	21	Milwaukee, Wis.
AT&SF	SD24	SD26	645E	2650	16	60'8"	35'0"	2/73-3/78	80	San Bernardino, Calif.
ICG	SD24, SD35	SD20	645E	2000	16	60'8"	35'0"	7/79-12/82	42	Paducah, Ky.
B&O	SD35	SD20-2[3]	645E	2000	16	60'8"	35'0"	7/79-1980	5	Cumberland, Md.
SP	SD35	SD35R[3]	645E	2000	16	60'8"	35'0"	10/76-1977	10	Sacramento, Calif.
SCL	SD35, SDP35	H15[3]	567BC	1500	16	60'8"	35'0"	6/77-9/78	8	Waycross, Ga.
L&N	SD35	SD35M	645E	2000	16	60'8"	35'0"	1979	2	South Louisville, Ky.

[1]These are unofficial model designations by the railroad, except where the original designation has been retained.
[2]SD9 dimension; former SD7s are ½" longer.
[3]Deturbocharged.

SD24

Union Pacific 3999 (also numbered 3100, 3200, and 3399 at various times; its original number was 423) was considered a possible prototype for conversion of all UP's SD24s. As it turned out, the unit was of more interest as a test bed for the concepts of a constant-speed engine (power output was varied by excitation of the alternator) and a single-wire electrical system (no return wires; all circuits used the chassis as ground through appropriate resistances). The constant-speed portion of the experiment was soon concluded and was not duplicated. The locomotive remained in service as an "almost SD45" and was later renumbered 99. Photo by Keith E. Ardinger.

SD24-4

Number 414 was UP's next re-built SD24, using latter-day (but non-modular) electrical components and an AR10 main alternator in place of the original DC generator. It, too, failed to become the prototype of a general program. UP wound up selling most of its SD24s to Precision; 20 of them became seeds for ICG's SD20 program. Photo by George Cockle; collection of Kenneth M. Ardinger.

Milwaukee Road 559 is typical of the SD10s, which are characterized by chopped noses and hooded paper air-filter intakes. Number 559 has an SD7 frame, as indicated by the round handrail stanchions. Soo Line (successor to Milwaukee Road) has leased most of the SD10s to the Minnesota, Dakota & Eastern. Photo by Kenneth M. Ardinger.

SD10

Santa Fe 4600 represents that road's extensive rebuilding of its SD24s. The installation of a central air intake behind the cab displaced the four rooftop air reservoirs to positions farther back along the edge of the hood, making the rebuilds easy to spot. Interestingly, these units accepted air conditioners without resorting to the "Topeka Cab." All the original 80 were gone by mid-1988, mostly as trades on GP50s, but some went to Guilford. Photo by J. R. Quinn.

SD26

SD20

Illinois Central Gulf based its SD20 program on former Union Pacific and Southern Railway SD24s obtained through Precision. Most of the UP units were SD24Bs to which it was necessary to add cabs; all of the Southern units had high noses which had to be lowered. Number 2027 was rebuilt from Southern 6327. Photo by Greg Sommers.

SD20

The last four SD20s, 2038-2041, were rebuilt from Baltimore & Ohio SD35s 7438, 7439, 7440, and 7437. Number 2038 illustrates this last batch. Photo by George Cockle.

SD20-2

In addition to sending four SD35s to ICG's SD20 program, Baltimore & Ohio carried out a similar deturbocharging operation at its Cumberland, Maryland, shop. SD35s 7403, 7405, 7401, 7404, and 7406 became SD20-2s 7700-7704. They are now CSX 2400-2404, and they still work in the Queensgate (Cincinnati) hump yard service for which they were built. Cumberland stuck with the conventional two-stack design as on an SD38, rather than resort to the four-stack manifold of the ICG rebuilds, and also removed the center radiator fan, which ICG retained. Photo by Greg Sommers.

Southern Pacific rebuilt ten SD35s at its Sacramento shop between 1977 and 1980, deturbocharging and rerating them to 2000 h.p. in the process. They are numbered 2961-2970, classed ES620 (EMD, switcher, 6 motors, 2000 h.p.), and used in hump service. The air filter box signifies a nonturbocharged unit, and in this view of 2962, the two exhaust stacks are barely visible fore and aft of the two central dynamic brake cooling fans. All three SD35 radiator fans have been retained. Photo by Bryan Griebenow.

H15

Perhaps the most peculiar SD35 conversions are Seaboard Coast Line H15s 1-8 (now CSX 1001-1008). They had their original turbocharged 2500 h.p. 567D engines replaced by nonturbocharged 1500 h.p. 567BC engines, presumably salvaged from GP7s, F3s, or F7s. They are used as hump power at Waycross, Georgia. The locomotives converted for this group were:

SDP35 1957 to 1	SD35 1901 to 3	SDP35 1951 to 5	SD35 1917 to 7
SDP35 1955 to 2	SD35 1916 to 4	SDP35 1964 to 6	SD35 1908 to 8

SDP35s had the long hood shortened 42 inches behind the radiator to allow for more end platform, and all the H15s were equipped with manual transition in four stages — two series-parallel stages and two shunt stages — presumably for better control in humping, although transition is usually not made in the very low speed ranges used in humping. Photo by Greg Sommers.

Louisville & Nashville's South Louisville Shop deturbocharged two SD35s in 1979 for the Louisville hump yard, where they still work for successor CSX. The illustration shows that the original radiator fans were retained and that two exhaust stacks and an air filter box replaced the single turbocharged stack. The locomotive is similar to SP's rebuilt SD35s. Photo by Louis A. Marre.

Chicago & North Western's "Oelwein rebuild" SD7s and SD9s 6601-6621 are not tabulated as they are simply over-hauls in kind in numerical sequence of 1701-10, 1721-24, 1660-64, and ex-M&StL 300 and 301, all done in mid-1971. All except the 1660-64 group originally had dynamic brakes, which were removed. Numbered above these units are "SD18s" 6622-6647, SD24s that were deturbocharged by Precision in 1979 and 1980 before sale to C&NW in 1981. They are included in the contractor section of the book. Both photos by Jack Armstrong.

Southern Pacific's extensive SD9E program is not tabulated, as it is primarily an overhaul in kind, except for **SD9E**
the installation of central air filters. Photo by Vic Reyna.

GE ROAD-SWITCHERS, REBUILT BY RAILROAD, 3100 H.P., C-C, AND 3000 H.P., B-B

In May 1985 Santa Fe initiated a significant capital rebuild program at Cleburne, Texas: the SF30C. The initial unit, 9500, was rebuilt from wrecked U36C 8721, and 96 more Santa Fe U36Cs were to follow.

The locomotives were downrated to 3100 h.p. and were given Dash-8 electronics, including the Sentry wheelslip system. Electrical modules were moved from under the cab to behind it (to a space provided in the original design for a steam generator), and Dash-7 improvements were made, such as the tilted oil cooler to permit complete draining in winter shutdowns, avoiding freeze damage.

This program is the first railroad effort to rebuild and update batches of the the six-axle, 3000-3600 h.p. models that predominated in the massive buying programs of the 1970s. As the market is glutted with lease turnbacks from this period, and as the price of new locomotives of this type reaches $1.5 million or more, there will be more such rebuildings, both by railroads and by contract rebuilding shops, perhaps to the detriment of locomotive manufacturers. The SF30C program was stopped after 70 units. The remaining unrebuilt U36Cs continue in service.

Number 9518 exhibits the muffled stack and the cab air conditioner. Photo by Greg Sommers.

The angle of this photo of 9511 shows the Dash 8 style nose and the hood bulge for the tilted oil cooler just in front of the radiator — features which distinguish the SF30C from the U36C. Photo by John B. McCall.

In mid-1987 as the SF30C program was winding down, Cleburne Shops produced a single prototype for a similar Capital Rebuild Program that could use the 34 remaining U23Bs as fodder. SF30B 7200 differed in several respects from the SF30C project: It was a significant uprating, from 2300 to 3000 h.p.; it involved extensive modifications to the hoods, principally to obtain sufficient area for the enlarged cooling system; and it included an extraordinary fuel tank which projected above the frame on each side — large enough for a Chicago-Los Angeles trip without refueling. This interesting experiment was not repeated by press time, and Santa Fe closed the Cleburne Shops at the end of 1987, leaving only San Bernardino capable of Capital Rebuild Programs. Late in 1988, San Bernardino started a limited upgrading program on the U23Bs, raising horsepower to 3000, but not altering the hoods or tanks as drastically as with the SF30B. Both photos by John B. McCall.

REPOWERING WITH CATERPILLAR ENGINES BY RAILROAD SHOPS

Caterpillar diesel engines were used frequently in the 1940s and 1950s in such locomotives as the GE 44-ton-ner. The first installation of a Caterpillar engine in a large locomotive was accomplished by Grand Trunk Western, which repowered Alco S-4 No. 8082 with a 975 h.p. V12 engine with 6¼″ × 8″ cylinders in November 1979. The repowered unit was renumbered 1000 and classed CS-9. Subsequently, 1001 and 1002 were created from 8084 and 8199 in 1980, and 1003 from 8162 in 1981. Photo by Louis A. Marre.

Caterpillar's 3500 series engine has 6.7″ × 7.5″ cylinders. Norfolk Southern used model 3512, the V12 configuration, which has a rating of 1050 h.p. at 1800 rpm, to repower five GP9s into TC10 transfer locomotives. TC10 100 was created from 518 in 1984, 101 and 102 were created from 519 and 520 in 1985, and 103 and 104 were created from 513 and 508 in 1986. These are non-turbocharged engines exhausting through a single stack concealed between the carbody filter projections on each side of the hood. Although NS was reported satisfied with the TC10s, it is not known how long the program will continue. Photo by Greg Sommers.

Of greater significance is Caterpillar's 3612 engine, an 11″ × 11.8″, 3800 h.p. V12. Chicago & North Western used such an engine to create SDCAT 6000 from Burlington Northern SD45 No. 6547 in February 1986. There is a separate turbocharger for each bank of cylinders on the 3612 engine, resulting in two stacks side by side — concealed behind noise baffles. BN used SD40-2 6330 as the basis for a similar conversion at West Burlington in December 1986, but retrofitted SD45 radiators to handle the larger cooling requirements of the 3800 h.p. rating. Photo by Bob Baker.

Santa Fe was the third experimenter, repowering SD45-2 5625 at San Bernardino in July 1987 to create SD CAT 5855. A similar CSX project was canceled, leaving three of the Big Cats loose in the West on AT&SF, BN, and C&NW. Photo by Brian Matsumoto.

REPOWERING BY RAILROAD SHOPS

Repowering is a form of capital rebuilding. The repowering programs of the 1950s and 1960s, most of which involved using EMD engines in non-EMD locomotives, were precursors of the more recent EMD-for-EMD capital rebuild programs. The early repowering era is documented in THE SECOND DIESEL SPOTTER'S GUIDE, pages 433-454. Here we illustrate some repowerings that have occurred since 1972.

Railroads are not repowering as many units as they did in the 1950s and 1960s for several reasons. Many candidates for repowering have been traded in instead. It is difficult to obtain new engines from EMD. Some earlier repowerings were not successful because auxiliary systems were outdated or worn out. Railroad shops have difficulty competing economically with outside contractors.

Repowering of Alco Road-Switchers, 1200-2000 H.P., B-B

Railroad	Model	Engine	H.P.	Cyls.	Length	Truck centers	Period produced	Approximate number of units produced	Shop location
PC, CR	RS-3	567B, C	1200	12	56'6"	30'0"	1972-1978	55	Altoona, Pennsylvania, and Syracuse, New York
Amtrak	RS-3	567B	1200	12	56'6"	30'0"	1981-1984	3	New Haven, Connecticut
GB&W	RS-3	251C	2000	12	56'6"	30'0"	1975	4	Green Bay, Wisconsin
L&N	RS-3	251C	2000	12	56'6"	30'0"	1973	1	Louisville, Kentucky
LSSCo	RS-2	KTA3087	1800	12	56'0"	30'0"	1982	1	Lone Star, Texas

The first unit in Penn Central's program to place E-unit engines in RS-3s was this unique locomotive with the short hood eliminated entirely, formerly PC 5477. The locomotive was assigned to yard service, so the lack of a short hood was sensible. The subsequent 54 units retained their short hoods. Doors had to be cut into the curve at the top of the hood because EMD power assemblies must be lifted higher than those of the Alco engine during replacement. Conrail continued the repowering program it inherited from PC. Photo by Don Dover.

Conrail 9994 represents the 1978 Altoona program, in which the power-assembly access doors continued the pattern set by nose-less 9950. Note the use of two EMD cooling fans but retention of the Alco radiators. The Altoona and De Witt repowerings were numbered from 9945 to 9999 at the end of 1978. Photo by Herbert H. Harwood.

To obtain the clearance needed for lifting out power assemblies, De Witt shops (Syracuse) created a boxy structure on top of the hood, as shown on No. 9954. Photo by Herbert H. Harwood.

Amtrak emulated the De Witt pattern in repowering three RS-3s at its New Haven Shop between 1981 and 1984. They were given 12-cylinder 567 engines salvaged from retired E units, and they were numbered 104, 106, and 107. The most visible departures from Conrail practice are the single exhaust stack and the neatly lowered short hood. What might have been a continuing program ended when Amtrak acquired 25 CF7s and 18 SSB1200s from Santa Fe in exchange for 18 SDP40Fs in September 1984. Photo by Gordon Lloyd Jr.

Green Bay & Western and Louisville & Nashville repowered RS-3s with Alco 251 engines, which required raising the hood slightly (as on GB&W 305-308) or providing a raised area over the turbocharger (as shown on L&N 1350, the road's sole repowered RS-3). These repowerings are in addition to contract repowering of RS-3s by Morrison-Knudsen for Delaware & Hudson and Detroit & Mackinac described later. Both photos by Louis A. Marre.

Lone Star Steel's Texas & Northern Railroad has had an extensive program of repowering 539-engine Alcos, about which we have few details. RS-1 43 is typical, with a small, single-stack engine sitting in the former radiator compartment of ex-Gulf, Mobile & Ohio 1054. The engine appears to be of a size that would produce 300-400 h.p. More ambitious is No. 996, which uses the frame of GM&O RS-2 1522 and a Cummins KTA3067 engine rated at 1800 h.p. It was outshopped May 10, 1982, and serves as a mother unit for slug No. 3, built from Wabash S1 159. Both photos by Bryan Griebenow.

REPOWERING OF ALCO ROAD-SWITCHERS BY RAILROAD SHOP, 1200-2050 H.P., C-C

Railroad	Old Model	New Model	Engine	H.P.	Cyls.	Length	Truck centers	Period produced	Approx. number of units produced	Shop location
PC, CR	RSD-15		567B	1200	12	66'7"	43'6"	1/75	1	Syracuse, New York
AT&SF	RSD-15	CRSD20	645E	2050	16	66'7"	43'6"	6/74-4/76	3	San Bernardino, California
FCP	RSD-5	API620, BX620	251C	2000	12	55'11¾"	34'9"	1979-1986	23	Empalme, Sonora

Conrail 6849 (ex-Penn Central 9949; earlier PC 6811 and Pennsylvania 8611) was permanently paired with an RSD5 slug of the same number. A plan to make a second pair was not carried out. The unit was assigned to De Witt hump yard at Syracuse, New York. Conrail retired the set in favor of SD38 and U23C hump power. Photo by Tom Trencansky.

In June 1974 Santa Fe created CRSD20 No. 3900 from Alco RSD-15 No. 9828 to work with a slug in hump duty. Nearly two years later, 3901 and 3902 were created from 9826 and 9846 in March and April of 1976. They were renumbered 1300-1302 in January 1977. They worked almost exclusively at the new hump yard at Barstow, California, until 1984, when they were sent to GE as trade-ins. Photo by Louis A. Marre.

Among the few substantial upgrades of Montreal Locomotive Works products are a series of rebuilt RSD-5s on Mexico's Ferrocarril del Pacifico. Empalme Shops obtained ''kits'' from Alco Products and Bombardier (successor to MLW) to upgrade 23 units between 1979 and 1986. The rebuilds are rated at 1800 h.p. for traction. Number 557 is one of 13 Alco versions, designated API620 (Alco Products Inc., 6 motors, 2000 gross h.p.). The Bombardier versions are BX620 (Bombardier Export, 6 motors, 2000 h.p. gross). Photo by Jim Herold.

EMD E9BS CONVERTED TO SUBURBAN-SERVICE CAB UNITS

From January through May 1973, Chicago & North Western converted 11 ex-Union Pacific E9Bs to "Crandall Cab" units at Oelwein shops for Chicago suburban service. The program included the installation of auxiliary engines for head-end power in the former steam generator compartment at the rear of the unit (the exhaust pipe and muffler are visible in the illustration). The "Crandall Cab" is named for its designer, C&NW Assistant Superintendent of Motive Power M. H. Crandall, who died of a heart attack while trying to restart a suburban locomotive during a blizzard in January 1979. Locomotive 502 was named in his honor. Photo by Lee Hastman.

CONTRACT REBUILDING

The economics involved in capital rebuilding were discussed in the introduction to the chapter on railroad shop capital rebuilds. However, some railroads either do not have the facilities to perform such work or find it advantageous to contract it out. The Precision National-Illinois Central Gulf team was the first and largest such endeavor, but it is no longer as active due to labor problems, a volatile market, and the sale of the Paducah shops. PNC has resumed contract work on a limited scale at Mount Vernon, Illinois.

Morrison-Knudsen of Boise, Idaho, was next to appear and remains active, though the frequency of orders has fluctuated during the 1980s. Chrome Locomotive (formerly Chrome Crankshaft) became a major force when it acquired the Silvis, Illinois, shops of the defunct Rock Island in 1980.

Contract capital rebuilding must be distinguished from mere overhaul or repair of units owned by the numerous equipment brokers and lessors. In order to qualify as capital rebuild, the price of the work done must exceed half the price of the original unit. Work on that scale requires extensive shops, something beyond the reach of most secondhand dealers. The distinction is not always sharp — in this book it is made primarily on mechanical grounds. Work that would be routine overhaul for a railroad shop is not considered capital rebuilding.

Contract shops continue to face the obstacle of railroad labor agreements, which often prevent contracting out shop work. Such objections can be overcome if the railroad can satisfy the unions that the units in question would otherwise be traded in or that such work is beyond the capabilities of its own shops.

CONTRACT REBUILDING: PRECISION NATIONAL CORPORATION

Precision Engineering Company began operation in 1931 as a rebuilder of diesel crankshafts. In 1960 it acquired Ford's Auto-Lite plant in Mount Vernon, Illinois, created a locomotive division, and became a broker, rebuilder, and short-term lessor of used locomotives. In 1969 it changed its name to Precision National Corporation, and in 1971 it entered a partnership with Illinois Central for rebuilding locomotives at IC's Paducah, Kentucky, shops.

For minor reconditioning work, PNC continued to use its own shop at Mount Vernon. The work on 40 Conrail units in 1979 was divided, 10 going to Mount Vernon and 30 to Paducah. The rebuilt units are generally called "Paducahs" regardless of where the work was accomplished.

PNC's work at Paducah ceased after completion of Clinchfield's GP11s in 1979. PNC is still a broker and lessor of used locomotives but no longer undertakes the volume of rebuilding it did in the 1970s.

PNC contract rebuilds generally resembled ICG rebuilds. However, customer specifications frequently differed from ICG's. Conrail 5407, for example, is one of 9 GP7s rebuilt at Paducah in 1976. At Conrail's request, the dynamic brakes were retained from the source unit, Erie-Lackawanna 1214, as was the original cab. The lower-profile single-stage Dynacell paper air-filter assembly ahead of the dynamic brake blister had come to be a feature of ICG rebuilds by the time the Conrail order was received. Photo by J. R. Quinn.

Ashley, Drew & Northern 1811 is a PNC-Paducah GP10, rebuilt from Illinois Central GP9 9201 in May 1979. AD&N's roster includes identical twin 1810, ex-IC GP9 9352. Low nose, Paducah cab, Dynacell filters, and four-stack exhaust are all earmarks of the Paducah GP10. Photo by Louis A. Marre.

FP10 is the informal model designation for 13 Paducah rebuilds of former Gulf, Mobile & Ohio F3As for Massachusetts Bay Transportation Authority. Numbers 1150-1153 retained their steam generators, but the remaining 9, including 1109 shown here, received head-end-power (HEP) diesel engine-alternator sets. All units in the program retained their 1500 h.p. 567B prime movers. Rebuilt in 1979, these were among the final units from Paducah. Photo by Gordon Lloyd Jr.

Conrail's contract rebuilding program was as follows. It is a good example of such a program.

PNC-Paducah
    GP8s 5400-5408 (1976, originally came from PNC as 5720-5728)
    GP8s 5409-5413, 5428, and 5429 (1978)
    GP10s 7545-7559, 7576-7597 (1978)
    GP10s 7560-7575 (1976)
PNC-Mount Vernon
    GP10s 7530, 7531, 7533-7535, 7537 (1978)
Rock Island-Silvis
    GP8s 5430-5449 (1978)
Morrison-Knudsen–Boise
    GP8s 5450-5462 (1978)
    GP10s 7513-7529 (1978)

Numbers 5414-5427 were left vacant for further GP8s. It may have been intended to fill the series 7500-7512 downward with future M-K units, although it was occupied by GP9s 7500-7508 at the time. The Mount Vernon series was to have been 7530-7539. The vacant numbers were intended for units that were rebuilt at Paducah instead, and 7540-7544 were intended for future Mount Vernon rebuilds.

Conrail GP10s 7566 and 7569 illustrate versions with and without dynamic braking, depending on the locomotive used as the basis for the rebuild. Number 7566 is ex-New York Central 5911; 7569 is ex-Pennsylvania 7118. Note that retention of the dynamic brake prohibits the four-stack liberated-exhaust manifold, and that the 7566, perhaps for uniformity through the contract, also has only two exhaust stacks. Both photos by Kenneth L. Douglas.

Alaska Railroad commissioned a prototype GP7 rebuild from contractor Morrison-Knudsen in 1975. Satisfied with the result, road number 1810, it ordered nine more from PNC-Paducah. These were delivered from April 1976 to April 1977, numbered 1801-1809. Number 1804 is the only one of the group to have dynamic braking. The other features are common to all ten: Alco type B road trucks (with roller bearings added) from scrapped Army RS-1s, low nose, permanent winterization cover over the radiator fan, and a cab like that of contemporary EMD production. The units have 645-series power assemblies and are rated at 1800 h.p. The truck replacement was dictated by the original switcher trucks on the GP7s, a peculiarity shared only with other ex-Army GP7s and the first order delivered to Nashville, Chattanooga & St. Louis, the latter long since scrapped. Photo by Curt Fortenberry.

The only GP11s built by Precision for a customer other than ICG were Clinchfield 4600-4605 (now CSX 1700-1705), delivered between March and May 1979. Subsequent rebuilds of Clinchfield GP7s were part of parent Seaboard Coast Line's Uceta Shops GP16 program. The GP11s were retroactively renamed GP16s in line with Uceta's nomenclature, and they appear on the CSX roster as such. They are different from Uceta GP16s, however, with a shortened nose, necessitated by moving the cab forward to accommodate a Dash 2 electrical cabinet, and the left side blower duct, both distinctive GP11 features. Photo by Louis A. Marre.

Perhaps the last contract rebuildings performed by the PNC-ICG partnership (and, we believe, at Mount Vernon, not Paducah) was the deturbocharging of ex-Southern and ex-Union Pacific SD24s to create PNC "SD18s." This work was done in 1980 and used up some of the stock that might otherwise have become ICG SD20s, compelling that program to substitute several B&O SD35s. Chicago & North Western acquired these SD18s in 1981 and put them through Oelwein Shops for further work, releasing them for service between August and December 1982. Former Southern high-short-hood units were numbered 6622-6643, and former UP low-nose units, 6644-6647. These units lack the Dash 2 electronics, Paducah cab, central air systems, and 645 power assemblies of their ICG SD20 counterparts. C&NW 6625 photo by George Cockle; C&NW 6646 photo by Roger Bee.

CONTRACT REBUILDING: MORRISON-KNUDSEN

After several years of maintaining locomotives (primarily Alcos) purchased secondhand for rail construction projects (such as Great Northern's Libby line change in 1969 and 1970), Morrison-Knudsen, of Boise, Idaho, entered the contract rebuilding field in 1973. M-K's first project was modification of 21 Burlington Northern E9As for the West Suburban Mass Transit District. The 2400 h.p. rating was retained, but 645 power assemblies were substituted. A 500 kw Detroit Diesel auxiliary power plant replaced the steam generator. Two ex-

tra cooling fans were installed on the roof of each unit between the exhaust stacks and are evident in the photo of BN 9903. In 1978, 4 E8As were rebuilt to the same specifications for WSMTD, bringing the total of BN-lettered and BN-maintained rebuilt E units to 25, all leased to WSMTD.

This contract was followed by M-K overhauls of several Amtrak E units, then by the 1975 upgrading of Delaware & Hudson's four PA-1s to 2400 h.p. with new 16-cylinder 251 engines.

The extra cooling fans and the exhaust mufflers for the auxiliary engine are visible atop Burlington Northern 9903. BN's rebuilt Es work commuter trains between Chicago and Aurora, Illinois. Photo by J. R. Quinn.

Delaware & Hudson's four rebuilt PA-1s were designated PA-4s. They were subsequently sold to Mexico and later retired. At press time No. 19 had been taken out of retirement and was being rebuilt at National Railways of Mexico's Empalme shops (ex-Ferrocarril del Pacifico). The rebuilding includes a Bombardier 12-cylinder 251 engine and new wiring. Photo by Jack Armstrong.

PA-4

Another of M-K's rebuilding programs was a standard package for converting an RS-3 to a 251-engined, low-nose locomotive of 1800-2000 h.p. The first of these was 1800 h.p. Detroit & Mackinac 974 rebuilt from Boston & Maine RS-3 1512 in December 1974. Next came D&M 975, a 2000 h.p. version rebuilt from B&M RS-3 1517 in 1975.

Delaware & Hudson 501-508 were rebuilt from D&H RS-3s 4115, 4106, 4107, 4113, 4123, 4112, 4119, and 4128 respectively between December 1975 and March 1976. To accommodate the model 251 engine, M-K had to raise the long hood 6 inches and displace the dynamic brake from its original position above the engine to a box at the top of the hood behind the cab. Part of the modification package for the D&H units was a single traction motor and generator blower behind the cab. The blower duct is on top of the left-hand running board, like the duct arrangement used by EMD since the GP30.

External differences in the RS-3s rebuilt by Morrison-Knudsen for Delaware & Hudson are short hood designated as front, low nose, central air filtration, and blower duct along the left running board. Photo by Jack Armstrong.

Despite its familiarity with Alco diesels, with Alco out of business M-K found little interest in them on the part of railroads. Even while the Alco rebuildings were in progress, M-K purchased a group of Union Pacific U25Bs in 1974. Eight of these units were rebuilt for Weyerhaeuser Company in 1975 and 1976. UP 637 was repowered with the 16-cylinder 567B engine and D12 generator of a B&O F7 in December 1975 and became Weyerhaeuser 310. The remaining seven units were rebuilt for Weyerhaeuser's Oregon, California & Eastern Railway. OC&E 7601-7605 were virtually identical to No. 310 (EMD innards with the U25B carbody largely intact), while 7606 and 7607 became road slugs. Photo by Kenneth M. Ardinger.

Two ex-UP U25Bs received treatment different from the OC&E units, as illustrated by 5302. The U25B underframe is barely recognizable. The cab is M-K's adaptation of the Canadian Comfort Cab concept, except for a rolling door in the nose, and the hood features an adaptation of EMD radiators and radiator fans, plus sliding power assembly doors. Both have been scrapped. Photo by Louis A. Marre.

Another M-K repowering that used EMD components was Manufacturers Railway 253, formerly Alco S2 208, outshopped in August 1976 with a 12-567BC engine and EMD generator and cooling-system components. The unit was rated at 1200 h.p. Photo by Mike Wise.

Amtrak obtained 12 ex-New Haven FL9 B-A1A diesel-electric-electric locomotives from Penn Central. (Metro-North acquired 37, and the remaining 11 were scrapped by Conrail.) M-K rebuilt 6 of the Amtrak units — 485-489 and 491 — between May 1979 and November 1980. A seventh unit, No. 484, was cannibalized during the rebuilding program. Rebuilding was largely in kind, except that D77 traction motors from scrapped SDP40Fs were substituted for the original motors, which had been a weakness of the FL9 in straight electric operation. Number 488 returned from Boise with its steam generator intact, but it was replaced with an HEP unit at New Haven in January 1982. The other 5 units were fitted with HEP units either at Boise or upon return to Amtrak. Four of Amtrak's rebuilt FL9s remain in service; the other 8 have been scrapped. Metro-North's units are still operating. Four Metro-North units owned by the Connecticut Department of Transportation have been rebuilt by Chrome Locomotive and repainted in their original New Haven livery. Photo by Jack Armstrong.

In 1980 M-K rebuilt wrecked Southern Pacific SD45s 9505 and 8820 using the dual hoods of SP DD35 9900, and numbered them 8301 and 8302. SP SD40 8402 was rebuilt with a similarly squared-off hood and numbered 8303. The three units have operated primarily on the Missouri-Kansas-Texas in their Morrison-Knudsen colors. They were intended for service on a piece of the former Rock Island between Belleville and Manhattan, Kansas, which Morrison-Knudsen was operating under contract to the Kansas Port Authority. That operation ceased in November 1980 and a long-term lease to Missouri-Kansas-Texas followed. Sheet metal aside, there is believed to be nothing mechanically unusual about these rebuilds. Both photos by Louis A. Marre.

While the ATSF/SP/UP Sulzer campaign was still under way, M-K built a demonstrator to test the market for GP-series rebuilds. Union Pacific GP9 278 was repowered with a 6-cylinder Sulzer engine, model 6ASL 25/30 (25 and 30 are the bore and stroke in centimeters). It was completed in July 1979 and designated model TE50-4S. The locomotive toured a number of roads including several short lines — potential customers for low-horse-power units — and is shown here on the DeQueen & Eastern in 1982. No orders were forthcoming, and the model has not been repeated. Number 5001 was last reported to be the shop switcher at M-K's Hornell, N. Y. plant. Photo by Bob Graham.

While searching for a market niche, M-K became interested in repowering locomotives with alternatives to EMD prime movers. It entered a licensing agreement with Sulzer Bros. Ltd., a Swiss diesel engine manufacturer with licensees in France and Great Britain. Sulzer was also a pioneer in diesel railway traction. In 1978, M-K repowered four Southern Pacific U25Bs with Sulzer 8-cylinder in-line turbocharged 8ASL 25/30 engines rated at 2800 h.p. The rebuilt units were also equipped with imported unitized radiator systems. Neither of the exotic components held up under U. S. operating conditions, and the SP experimentals were soon retired. At press time, no Sulzer-powered locomotives are in common-carrier service in the U. S. M-K's model number system is tractive effort in thousands of pounds; number of powered axles; and a letter for prime mover builder. The repowered SP U25Bs were designated TE70-4S. Photo by Kenneth M. Ardinger.

The most significant repowerings in the Sulzer campaign were ten SD45s converted in 1980 as a joint effort of M-K and the shops of the Santa Fe and the Union Pacific. Santa Fe SD45s 5515, 5541, 5530, and 5551 became Sulzer-powered 5496-5499 in 1980 and 1981. The Santa Fe experiment lasted only until these units were converted back to 20-cylinder EMD engines in January and February of 1985, taking road numbers 5405-5408. Union Pacific converted six SD45s (34, 14, 15, 13, 8, and 37) between August 1980 and January 1982. The rebuilt units were numbered 60-65. The UP experiment ended with the retirement of all six units in December 1983. Both railroads performed the sheet-metal work, including widening the hood two inches on each side in the power-assembly door area, then sent the locomotives to Boise to have the engines installed. The engine used was V-16 design, model ASV 25/30, initially rated at 4000 h.p. but later reduced to 3600 h.p. M-K designated these units TE83-6S. Photo by Kenneth M. Ardinger.

Commuter operations are money losers, and the cost of new equipment has become increasingly difficult to justify. When the price of a new F40PH reached $2 million Morrison-Knudsen found a market for substitutes. At press time, M-K had built and delivered three orders of units that are similar in concept, albeit diverse in appearance. The most conventional in appearance are six Maryland Department of Transportation (MARC) units called GP39H-2s. (M-K appears to have abandoned its model number system.) These are GP40s with new 12-645E engines plus a Cummins HEP generator inside a slightly extended hood. The clearest indicator that these are not GP40s is the uneven spacing of the radiator grille at the rear. All six were built at Boise between November 1987 and March 1988. Photo by Herbert H. Harwood.

More ambitious F40 clones are 14 GP40FH-2 units for New Jersey Transit (10) and Metro-North (4). They were constructed at Boise in 1987 and 1988. A variety of secondhand GP40s provided the frames, while salvaged BN F45s provided 10 full-width cowls, supplemented by 4 copies constructed by M-K. The result is a 3000 h.p. unit with a new 16-645E3 prime mover plus a large 600 h.p. Cummins HEP set at the rear. The combination of F45 cowl and GP40 cab is unlikely to be confused with anything else. Photo by Jim Herold.

Much closer in appearance to an F40PH are the most recent M-K models, five F40PHL-2s built for Tri-Rail's Miami, Florida, commuter operation. The L in the model number likely stands for longer, as these are built on GP40 frames, 3' longer than an F40PH. The Tri-Rail units have 16-cylinder 645E3A prime movers, and are rated at 3200 h.p., but not all for traction, as they derive their head-end power from the prime mover, as do ''real'' F40PHs. Photo by Gordon B. Mott.

CONTRACT REBUILDING: GENERAL ELECTRIC

After Erie Lackawanna's operations were taken over by Conrail in 1976, General Electric acquired the former Erie locomotive shop at Hornell, New York, primarily for transit-car work. However, the capabilities of the shop soon allowed GE to offer contract rebuilding of locomotives. Given the GE electrical equipment of Alco locomotives, it was natural that Alcos would predominate in the firm's work — though one of the first jobs was reconditioning five Metro-North FL9s in 1979. (Metro-North is an agency of New York state and understandably prefers in-state vendors.)

After reconditioning ex-Lehigh & Hudson River C-420 No. 27 as Green Bay & Western 323 in December 1979, Hornell began an interesting series of rebuilds of C-424s. Pennsylvania 2415, Erie Lackawanna 2412 and 2415, and Reading 5204 retained their 2400 h.p. rating and emerged from the rebuilding as GB&W 319-322 between January and March 1980.

In May 1980 Delaware & Hudson 451-456 and 461-463 were outshopped with new 12-251 engines downrated to 2000 h.p. The work included extensive rewiring, rebuilt radiators, new air filtration, and other significant changes designed to reduce maintenance costs. Numbers 451-456 were originally EL 2401, 2406, 2407, and 2414 and Reading 5206 and 5207. Units 461-463 were originally EL 2412, 2405, and 2408, and later became Genesee & Wyoming 61-63.

Next were four in-kind rebuilds of C-420s. Louisville & Nashville 1385, 1316, and 1333 and Norfolk & Western 415 became Apache Railway 81-84. The program of derated C-424s resumed with PRR 2439, 2443, 2442, and 2441, which became Detroit & Mackinac 1280, 181, 281, and 381 (the numbers represent the dates they were outshopped).

The market for Conrail Alcos was soon satisfied, and GE returned to rebuilding EMD units: Southern Pacific (later Wellsville, Addison and Galeton) F7s 6443 and 365 emerged as HEP-equipped 1750 h.p. F9s 6690 and 6691 of the Port Authority of Allegheny County for commuter service at Pittsburgh, Pennsylvania. During the 1981-1982 recession, GE closed the Hornell facility and eventually sold it to Morrison-Knudsen.

GE can still rebuild locomotives at its Cleveland Apparatus Shop, as evidenced by the 1986-1987 conversion of three ex-EL, ex-Conrail U33Cs to 1800 h.p. Cummins-powered cabless remote-control units (not illustrated) for the An Tai Bao coal mine in China, a joint venture of the Chinese government and Island Creek Coal.

A typical GE-Hornell rebuilt and derated C-424, Detroit & Mackinac No. 1280 is numbered for the month and year it was rebuilt. Photo by Gordon Lloyd Jr.

Port Authority of Allegheny County 6691 pulls (and pushes) commuter trains between Pittsburgh and Versailles, Pennsylvania. Photo by Gordon Lloyd Jr.

CONTRACT REBUILDING: CHROME LOCOMOTIVE

Chrome Crankshaft, Inc., of Los Angeles, California, was a supplier of reconditioned locomotive parts for many years and fell naturally into the business of buying locomotives to "part out." From that it was a short step to buying locomotives for resale. Renaming the business Chrome Locomotive, the firm purchased the former Rock Island shop at Silvis (East Moline), Illinois, after Rock Island shut down in 1980. Since then Chrome has reconditioned a large number of locomotives (especially former Rock Island units) that it acquired

for resale. The company also performs contract rebuilding.

Chrome's most significant rebuilding so far is the reconditioning of four FL9s owned by the Connecticut Department of Transportation for use on Metro-North trains. FL9s 5005, 5026, 5049, and 5057 were painted in their original New Haven colors in the process and are now numbered 2002, 2006, 2019, and 2023, interspersed among Metro-North's own FL9s, some of which were rebuilt by GE at Hornell and others by Metro-North in its Harmon shops.

The FL9 carries third-rail shoes on both trucks. The locomotive was designed to draw power from the third rail in the tunnels leading to New York's subterranean stations and run as a conventional diesel electric in open air, eliminating the need to change engines at New Haven. Changes in ownership and operational patterns put long-distance trains back in the hands of electric locomotives west of New Haven, and the FL9s found employment on commuter and Amtrak trains that serve New York's Grand Central Terminal. The original New Haven livery was restored to four Connecticut-owned FL9s during rebuilding. Photo by Jack Armstrong.

CONTRACT REBUILDING: PEAKER SERVICES, INC.

Peaker Services, Inc., of Brighton, Michigan, derives both its name and its diesel experience from rebuilding the EMD 567 and 645 engines used in peaker utility plants, which provide supplemental electricity at peak periods. Many small electric utility systems have such diesel plants, often powered by EMD engines virtually identical to locomotive prime movers. It was natural for Peaker to move into reconditioning locomotives, primarily EMD switchers. In 1981 Peaker rebuilt Grand Trunk Western GP18 4952 for the now-defunct Southeastern Michigan Transportation Authority commuter operation between Pontiac and Detroit. In addition to general reconditioning, the unit was converted back to GP9 specifications (1750 h.p.), but Peaker chose to call the result a GP19. Photo by George C. Diebel.

CONTRACT REBUILDING: DIESEL-ELECTRIC SERVICE

Diesel-Electric Service, Inc., of St. Paul, Minnesota, operated for a time in the 1970s and early 1980s as a dealer and reconditioner of second-hand locomotives, primarily switchers. The company's major rebuilding project was the reconstruction of wrecked Soo Line GP9 2554. The frame, trucks, and engine A-frame of the wrecked GP9 were used, but with 645 power assemblies. The cab, hood, and fuel tank were taken from wrecked SP GP35 6649, with the radiator end shortened to leave just the two large fans, eliminating the small middle fan. The unit was given the model number GP22 (the average of GP9 and GP35, perhaps) and was placed in service in May 1979. It was not successful and was traded to EMD in 1984. Diesel-Electric Service, Inc. has gone out of business. Photo by William A. Raia.

CONTRACT REBUILDING: CANADIAN NATIONAL RAILWAYS

In 1978 Southern Pacific turned to outside contractors for in-kind rebuilding of 60 GP35s. Morrison-Knudsen rebuilt 34 and Canadian National's Point St. Charles shop in Montreal rebuilt 26 (6324-6337, 6345-6352, and 6355-6358). CN has continued to advertise Point St. Charles as a contract shop, but so far the work, other than the SP order, has continued to be CN's own programs. It is extremely difficult for a railroad shop paying railroad wages to compete for this type of work. Photo by Gordon Lloyd Jr.

CONTRACT REBUILDING: LOW-POWER CONVERSIONS OF ALCO 539 SWITCHERS

A number of industrial locomotive dealers and small shops, as well as GE's Cleveland Apparatus Shop, have produced conversions of Alco 539-engine switchers (S-1, S-2, S-3, and S-4) for low-speed industrial applications. The 539 engine and generator are removed (or left in place as ballast) and a small engine, usually a Cummins in the 150-400 h.p. range, is placed in the former radiator area with an appropriate generator. The cab is generally removed, since most of these locomotives are fitted for remote control. The builder's plate, which was on the cab side, is usually lost as well, so the histories of these locomotives cannot be traced. Given all this, is it any wonder that they are called Zombies? Photo by Max Connery.

LEASE TURNBACKS AND FLEET DOWNSIZING

The overwhelming locomotive event of the current decade has been the impact of massive lease turnbacks on the locomotive market. Until the 1980s, most locomotives released from service by the arrival of second- or third-generation power were scrapped after being traded in to the builder of the replacement units.

EMD in particular had a firm policy of scrapping trade-ins to avoid creating a secondhand market that would depress new locomotive sales. Trade-ins were never economical for the builders in themselves, but a scrapped trade-in was a unit that had to be replaced with a new one.

In the 1980s the combined effects of sharp contraction of the railroad industry and improved locomotive productivity released far more power than the trade-in market could absorb. Many of these locomotives were relatively new second-generation units of 3000 h.p. or more. Contributing to this flood of serviceable secondhand power was the 15-year leveraged lease, which had been a popular way of using investment tax credits in the early 1970s, when railroads were earning far more credits than their meager profits allowed them to use on their own. An outside leasing firm could buy the locomotive, apply the tax credits against its own profits, and lease the locomotive to the railroad for its tax life, 15 years.

Railroads noticed in turn that the end of the 15-year lease coincided with a locomotive's second heavy overhaul and asked, in effect, why spend money for a heavy overhaul of a unit that's fully depreciated? Consequently, locomotives released at the end of a 15-year lease come out on the market due for heavy overhaul. This severely limits the price such units can command in the secondhand market, and has led to a lot of short-term use of lease-turnback "junkers" by power-short railroads, without the units finding a permanent home.

Two major players in this field are Helm Leasing Corporation and Chrome Locomotive. Their approaches are different. Helm, without a shop, serves as a sort of commodities broker in locomotives. It buys lease turnback and other superannuated locomotives that can be leased profitably for a short term and usually relies on the lessee to keep the units running. Such units often appear in the livery of their previous owner, crudely painted over in places and with HL or HLCX initials in place of the former road name. The ideal outcome for Helm is that some short-term lessee or other customer decides that the units are worth purchasing and overhauling or perhaps leasing for a longer term, such as five years.

Chrome Locomotive, which has a shop at Silvis, Illinois, has tended to seek "value-added" sales, giving locomotives a light to moderate overhaul and painting them. (Chrome also buys locomotives for parts and buys, sells, and arranges sales of unrebuilt locomotives.) The idea of buy-and-rebuild transactions appeals to the contract rebuilders and has also been taken up by EMD. For such work to be profitable, though, a rebuilt GP40 must fetch a price in the $600,000 range. It isn't yet clear that the market will support that, with so many junkers able to be patched up for less — and the railroad industry still shrinking.

Chrome Locomotive overhauled and repainted seven Rock Island GP40s for GO Transit. The former F7B coupled behind the GP40 is a head-end-power car that provides electricity for heating, cooling, and lighting the passenger cars. Photo by Gordon Lloyd Jr.

Helm Leasing 3082 had its Conrail identity painted out before being leased to Grand Trunk Western. It retains its Conrail, Penn Central, and New York Central number. Photo by Louis A. Marre.

One of the larger secondhand power transactions of the 1980s was Chicago & North Western's equipping itself for the opening of its Western coal operation, using cast-off Conrail and BN SD-45s. The program began with the purchase of four Conrail units in 1982, now CNW 6558-6561, with some confusion in CNW records about their original Conrail numbers. By the end of 1983 C&NW had purchased 67 Conrail and 23 BN SD45s through Chrome Locomotive:

C&NW	Conrail	C&NW	Conrail
6500	6170	6534-6557*	6211-6234
6501-6507	6172-6178	*6547 received a Caterpillar	
6508-6511	6180-6183	engine, was renumbered 6000.	
6512-6521	6185-6194	6558-6561	unknown
6522-6524	6196-6198	6562-6566	6235-6239
6525-6531	6200-6206		BN
6532, 6533	6208, 6209	6567-6589	6448-6471

The Conrail units were former Pennsylvania units of the same number, except for 6235-6239, which were delivered to PRR successor Penn Central. Of the BN units, 6448-6456 were ex-Great Northern 418-426, representing GN's entire 1968 order; and 6457-6471 were ex-Chicago, Burlington & Quincy 516-530, the Q's only SD45s.

This transfer of entire order-blocks of units at the expiration of leases (exactly 15 years from delivery) illustrates well the lease-turnback phenomenon.

C&NW then went on to acquire 12 more ex-BN SD45s in 1985 through Helm: 6472-6474, 6476-6478, 6481, 6482, 6485, 6488, 6490, and 6491. All were from a group of 20 units ordered by Northern Pacific and delivered to BN in 1970 — and all released at the end of their 15-year lease.

C&NW originally planned to put its 1982-1983 acquisitions through a rebuild program, for which 6500 became the prototype. It was given a general overhaul, Dash 2 modular electrical system, and Vapor PTC (Positive Traction Control), plus a switch to permit changing to EMD WS-10 wheelslip control.

C&NW did not carry through its proposed rebuild program, but instead concentrated its dynamic-brake-equipped SD40-2s in coal service and used the unrebuilt SD45s in general service. In 1986 C&NW traded in 17 of the unrebuilt SD45s on SD60s: 6551, 6552, 6554, 6555, 6558-6563, 6569, 6572, 6575, 6577, 6578, 6587, and 6588. C&NW had in the meantime acquired SD50s, which together with the SD60s obviated the need for a large-scale rebuild program.

Chicago & North Western 5525 began life in November 1965 as New York Central 3036, the first production GP40 and the first production 645-engine locomotive. It was not a lease turnback; its 15-year tax life expired in 1980 and, as a high-cost class, it was caught in the downsizing of Conrail's fleet. C&NW bought 38 Conrail GP40s — 3010-3023, 3025-3042, and 3044-3049. Rebuilding at Oelwein, Iowa, included engine overhaul, partial replacement of electrical cabinets with Dash-2 type, extended-range dynamic braking, and, for some units, Canadian National's PTC wheel slip control. The rebuilt units are labeled GP40-1.5. Photo by William S. Kuba.

C&NW 6564 was outshopped in February 1986 as an "SD40-2," with a 16-cylinder 645-E3B engine and Dash 2 components. Similar conversions have been done by Santa Fe and Southern Pacific. Photo by George Cockle.

Chicago & North Western SD45s 6530 and
6503 are veterans of Conrail, Penn Central,
and the Pennsy. Photo by Louis A. Marre.

Burlington Northern SD45 6452 received minimal repainting when it was acquired by the
C&NW. Photo by Louis A. Marre.

In December 1983 C&NW tested rebuilt SD45 6500 between St. Francis and Butler, Wisconsin, using two SD40-2s as a braking load. C&NW decided the PTC was not noticeably better than EMD's WS-10, but did note that the PTC control is designed for the SD40-2, not for the SD45. Photo by Jerrold F. Hilton.

Burlington Northern acquired 35 ex-Conrail GP38s in 1985, all but 8 of a group delivered in 1970 as Penn Central 7825-7867. Conrail 7825-7827, 7829-7836, 7838-7842, 7845-7848, 7850-7855, 7858-7861, and 7863-7867 became BN 2155-2189. Photo by Greg Sommers.

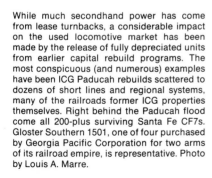

While much secondhand power has come from lease turnbacks, a considerable impact on the used locomotive market has been made by the release of fully depreciated units from earlier capital rebuild programs. The most conspicuous (and numerous) examples have been ICG Paducah rebuilds scattered to dozens of short lines and regional systems, many of the railroads former ICG properties themselves. Right behind the Paducah flood come all 200-plus surviving Santa Fe CF7s. Gloster Southern 1501, one of four purchased by Georgia Pacific Corporation for two arms of its railroad empire, is representative. Photo by Louis A. Marre.

FLEET LEASING

Burlington Northern leased 100 units, numbered 9000-9099, from Oakway Leasing under a "power-by-the-hour" contract. The terms of the lease are not as radical as the designation suggests. There is a periodic minimum usage which must be met on a take-or-pay basis, and that minimum is set close to the estimated average mileage of the units. The real purpose of the lease is to avoid costly shop labor agreements by having the units maintained through some arrangement with the lessor. BN has also taken delivery of 100 General Electric B39-8s, 8500-8599, owned by LMX Leasing. The Oakway units are maintained at a former BN shop at North Kansas City, and the LMX units at BN's Lincoln, Nebraska, shop, where GE personnel are assigned to work on them.

The B39-8Es that BN has leased from LMX are gray and striped in white and red. On the nose is a small GE monogram. Photo by George Cockle.

Eight blue-and-white Oakway Leasing SD60s in Electro Motive livery head west out of Lincoln, Nebraska, for Alliance on December 17, 1986, ready to enter coal train service under a new "power-by-the-hour" contract. Photo by George Cockle.

NON-POWERED CONVERSIONS OF LOCOMOTIVES

Over the years, a number of locomotive carbodies have been converted to non-powered uses, including cab-control cars and head-end power plants for push-pull commuter trains, heater cars, radio remote-control cars, and even fuel tenders.

Carbody-type units (usually old F units or similar units of other builders) are better than hood units for such conversions because there's more room inside. It is usually not economical to convert surplus non-carbody locomotives because extensive new construction is required and the savings resulting from the use of a ready-made carbody are lost. Also, many of these conversions were undertaken just as F units were reaching the ends of their economic lives. It was cheaper to convert an existing carbody locomotive than to build new cars for the purpose.

HEATER CARS

Passenger cars were originally heated by stoves, a ready source of ignition in case of an accident. In the 1880s stoves began to give way to steam piped from the locomotive. In 1890 the Milwaukee Road built a pair of electric-light-and-heat tenders for the *Pioneer Limited*. The 4-4-0s that pulled the train did not have sufficient steaming capacity to heat its cars, and generators had not developed to the point that they could be hung underneath each car to operate automatically. The heat tenders carried a coal-fired boiler and a steam-driven dynamo.

Great Northern heater cars 1-4 were built in 1928 by St. Louis Car Company. They were about 40 feet long and resembled short baggage cars. They were used behind electric locomotives, which lacked boilers, through the Cascade Tunnel and in later years on the rear of passenger trains across Montana, North Dakota, and Minnesota during extreme cold weather. At 30 or 40 degrees below zero, heat escapes from passenger cars at a rate which exceeds the ability of the steam line to supply the cars toward the rear of the train. By placing a second source of steam at the rear of the train and opening the steam line in the middle of the train to exhaust condensate, this problem can be overcome, and trains of normal length can be operated instead of running them in sections.

GN built more such cars in the 1940s, using passenger and box car bodies, then between 1965 and 1967 supplanted the older heater cars with 10 semi-automated cars which were less labor intensive — the old cars required full-time attendants.

Cars 10-19 were built from F3Bs and one F9B. The cars included lights which indicated whether the boilers were running or shut down. They cycled from one boiler to the other at specified intervals, and they blew down automatically. Equipment included two 4,500-pound-per-hour steam generators, two GM 220-volt diesel generator sets to run pumps, a 12,000-gallon water tank, and a 1,200-gallon fuel tank. The cars had M.U. capability and could be run in a locomotive consist, but were intended to be used at the rear of the train.

When Amtrak began operation, it purchased cars 10-15 and renumbered them 1910-1915. BN retained cars 16-19 for heating special trains, though Amtrak often rented them. With the conversion to electric heating of Amtrak's passenger cars and BN's business car fleet, the heater cars became surplus and were retired.

Amtrak's last regular operation of steam-heated cars occurred March 9, 1982, on the Florida-New York *Silver Star*. VIA Rail Canada uses heater cars occasionally, but they are ex-Canadian National freight-car-style cars, not converted locomotives.

The last regular use of a steam generator car in the U. S. of which the authors are aware was the Denver & Rio Grande Western's ski train between Denver and Winter Park. One of D&RGW's heater cars was a converted Alco PB-1 riding on EMD Blomberg trucks, and the other was converted from a steam locomotive tender. The ancient former Northern Pacific coaches which make up the train were replaced in 1988 with electrically heated Tempo cars purchased from VIA Rail Canada.

The F3B ancestry of Amtrak heater car 1915 is obvious. Photo by Kenneth M. Ardinger.

Water and fuel fillers, louvers next to the door, and smooth side panels indicate to the ground-level observer that Amtrak 666 was something other than an E9B — a heater car. Lack of cooling fans on the roof was another clue. Photo by Herbert H. Harwood.

Heater car 675, converted from an E8A, looked somewhat less finished, with grilles missing and cab windows blanked; roof details were retained. Photo by George H. Drury.

Alaska Railroad operated three converted E8B units for train heating and lighting: HEP-1 (later P-30) was rebuilt at Paducah from Illinois Central 4108 in September 1981, and steam generator cars P-6 and P-7. All three are former Amtrak units. The HEP generator car can be identified by the two pairs of exhaust silencers on the roof for the diesels that drive the 480-volt generators. The Alaska Railroad equipped F7B 1503 and E9A 2402 for HEP, but they remained locomotives; the cars shown have neither prime movers nor traction motors. Both photos by Curt Fortenberry.

HEAD-END-POWER CARS

Head-end power for lighting passenger cars is not a new idea — witness the Milwaukee Road electric light and heat tender mentioned a few pages back. However, the standard method for lighting passenger cars came to consist of an axle-driven generator and storage batteries underneath each car. Head-end lighting continued to be used on suburban trains, where low speeds and short runs prevented the generators from charging the batteries. Such systems were fed first from a large turbogenerator mounted on the locomotive boiler or the tender, then later from a diesel-driven generator in the locomotive or carried in a specially equipped coach or combine.

Some railroads decided it was cheaper to house the head-end power source in a locomotive carbody which could also serve as the control cab for the non-locomotive end of a push-pull train. The Long Island Rail Road and GO Transit use such units.

Any conversion process includes a transition period. As Amtrak converted its passenger car fleet from steam heat and axle-powered light to head-end power and replaced the SDP40Fs with F40PHs, it compensated for mismatches between locomotive and car by equipping several conventional coaches with HEP generators for Amfleet trains. Conversely, E units were rebuilt to heater cars for trains of conventional equipment.

E8s 495-499 had their steam generators replaced by HEP units at ICG's Paducah shops in 1975. That same year Paducah converted six E9Bs to heater cars 1916-1921 (later renumbered 666-671). In 1977 five E8As were converted to similar cars at Amtrak's Hialeah shop at Miami, Florida, becoming 1922-1926 (later 672-676). Amtrak retired its heater cars at the end of the transition period.

Amtrak E8s 495-499 were given head-end power apparatus in 1975 so they could be assigned to trains made up of new Amcoaches. The extra rooftop cooling fans flanking the stacks are for the HEP diesel. Photo by Kenneth L. Douglas.

275

FUEL TENDERS

Northern Pacific's "water baggage cars" provided a precedent of sorts for carbody fuel tenders. Space for water tanks aboard NP's F units was limited by boilers, dynamic braking grids, and extra-capacity fuel tanks, so the boiler water supply was carried in the adjacent baggage car and piped (trainlined) to the locomotives.

The small fuel tanks of the F40PH led to excess fuel stops in transcontinental service — undesirable from a scheduling standpoint. In addition, if Amtrak could eliminate fuel stops, it could reduce payments to contracting railroads for the maintenance of fuel facilities and their EPA-mandated track pans and skimming plants. Accord-

ingly, in 1978 E8A 400 was experimentally converted to a fuel tender. In its carbody were placed six 1350-gallon fuel tanks (former SDP40F water tanks) which added to the unit's regular fuel tank gave a capacity of 9300 gallons, enough for a Chicago-Seattle trip without refueling (a fuel tender requires that locomotives have trainlined fuel lines). The fuel tender was not duplicated and the experiment soon ended. Ironically, Burlington Northern, on whose lines Amtrak tested its fuel tender, now uses fuel tenders extensively in freight service; they are discussed a few pages farther on.

Amtrak 400's engines were removed and replaced with fuel tanks. Photo by George Cockle, collection of Kenneth L. Douglas.

PUSH-PULL POWER CARS AND CAB CONTROL CARS

Push-pull commuter trains had been in use in Europe for many years when the concept was introduced in the U. S. by the Chicago & North Western. In 1959 C&NW took delivery of a group of bilevel suburban coaches equipped with control cabs for pushing movements, thereby eliminating the need for locomotives to be uncoupled, turned, and moved to the other end of the train at terminals. The push-pull principle was copied by three other Chicago railroads for suburban service — Milwaukee Road in 1961 and Burlington and Rock Island in 1965 — and in 1967 by the Central Railroad of New Jersey, which added control cabs to old cars, and GO Transit, with new cars.

The Long Island Rail Road and its owner, the Metropolitan Transportation Authority, came up with a different idea in 1970 to avoid buying specialized commuter locomotives such as the GP40P and the GP40TC. LIRR acquired retired Alco FA-1s and FA-2s from various roads and converted them to control cab cars which also housed a head-end electric heating and lighting plant. It was discovered that a new engine wasn't necessary — the Alco 244 engine was derated to 600 h.p. and coupled to a new generator. Power cars 600-618, converted between 1970 and 1974, were joined in 1979 by 619 and 620, former Milwaukee Road F9A and F7A, respectively.

GO Transit of Toronto has kept a close eye on the Long Island, and in 1973 and 1974 converted five ex-Ontario Northland FP7s to similar power-cab cars. However, GO used a 900 h.p. GM model 149 engine instead of a derated prime mover to drive the alternator. Like the Long Island, GO Transit found an economical solution to its motive power problem: a conventional locomotive at one end of the train and a control cab and hotel power source at the other.

Long Island 601 is a former FA-2. Photo by Charles Trapani Jr.

Long Island 614 was converted from an FA-1. Photo by W. J. Brennan.

GO power-cab car 911, converted from Milwaukee Road FP7 No. 104A, trails a train led by a GP40-2. In 1976 the original units, 9858-9862, were renumbered 900-904 and four more ONR FP7s were similarly converted to power-cab cars 905-908. Two Milwaukee Road FP7s were similarly converted in 1982 and number 910 and 911. Also in 1982, GO purchased seven ex-Rock Island GP40s (GO 720-726) and seven ex-Burlington Northern F7Bs, which became 900 h.p. power cars 800-806. The HEP F7Bs are operated next to the GP40s, while the cab at the other end of the train is provided by a Hawker-Siddeley self-propelled diesel car that has been stripped of its engines. At times, another GP40 is operated at the other end of the train in lieu of these cab cars, but of course one end of the train or the other must have the HEP car. GO's new F59PH locomotives return to the original GP40TC concept of having the HEP in the locomotive, so the future of GO Transit's HEP cars is unclear. Both photos by Gordon Lloyd Jr.

In 1979 the Long Island created five more power cab cars: 619 from Milwaukee Road F9A 126A, 620 from MILW F7A 85A, 621 from Baltimore & Ohio F7A 4599, 622 from B&O F7A 4524, and 623 from B&O F7A 4535. Detroit Diesel 12-71 engines were installed to replace the old model 567 prime movers. FA power cab cars 605 and 606 were similarly converted to 12-71 engines in 1981 and 1982, and in addition MP15ACs 169-172 were converted so that they could provide hotel power at the flip of a switch, allowing them to stand in for ailing FAs. LIRR has begun to retire its FA power cab cars. Photo by Charles Trapani.

ROTARY SNOWPLOW POWER UNITS

A few old F units, both A units and B units, have found employment as power cars for rotary snowplows. The current generated for traction is rerouted to the rotary snowplow, whose blade is turned by traction motors identical to those that drive locomotives. This is not a major change in function for the locomotive, but simply a matter of additional wiring. Some rotary-power units have retained their traction motors, and some of Burlington Northern's units even went back into revenue service (with their maintenance-of-way numbers and brown paint) to ease a power shortage. In the long run, however, there is little point in maintaining traction motors under these units, as the plow is always pushed by several regular locomotives.

Milwaukee Road rotary power car X-2 is a former F7. Photo by Larry Russell.

An example of minimal reuse of a locomotive body is this snowplow conversion of dead Toledo, Peoria & Western RS-11 No. 402. Most roads are content to achieve this result by putting rocks or broken concrete in a gondola car to which a plow is attached — rocks have a scrap value far less than $70 per ton. Photo by Gordon Lloyd Jr.

RADIO CONTROL CARS

When Radiation, Inc.'s, Locotrol became the accepted means of controlling unmanned locomotives, two methods of installing the equipment became common. One was to use a radio control car that could control any locomotives attached to it, and the other was to install the Locotrol equipment in the locomotive itself, often in a specially lengthened short hood.

Old carbody locomotives are ideally suited for conversion to radio control cars because the necessary M. U. cables and control equipment are already at hand. The radio control equipment is installed in an insulated compartment inside the unit and takes up little space. Roads having such cars have included Burlington Northern, Louis-ville & Nashville, Chesapeake & Ohio, Santa Fe (whose fleet of 26, numbered 10-35, is one of the largest), Canadian Pacific, and British Columbia Railway (in Canada they are called robot cars). The largest fleet of radio control cars belongs to Southern Railway; they were built for the purpose instead of converted from B units.

Most railroads have decided to use only certain classes of locomotives as remote units, because successful operation depends on predictable responses from the remote units — something a mixed bag of engines cannot always provide. Radio control cars became just another piece of equipment to be maintained and they have begun to disappear.

Santa Fe's large fleet of radio control cars is being retired, and radio control equipment is being installed directly on the remote locomotives. Photo by J. R. Quinn.

Canadian Pacific Robot-5 is a remote control car with more than a soupçon of Alco-Montreal FB in its recipe.
Photo by Jim Herold.

AIR REPEATER CARS

Burlington Northern inherited from Great Northern eight air repeater cars which had been built from box cars between 1964 and 1970. During extreme cold weather (which causes increased leakage from air hoses) such cars help maintain air pressure in a long train. A small diesel engine driving an air compressor kicks in whenever brake pipe pressure drops below a set value.

During a spell when company management was assumed by men who weren't intimate with Northern Plains winters, the cars were discarded. Later, repenting of this, the company plated over some of the cab glass on U28B and U30B locomotives 5453-5459, 5773, and 5781 and tried to use them the same way — an idling, otherwise-unmodified locomotive coupled in a train will respond the same way as an air repeater car. However, untended locomotives, especially old GEs, have a way of shutting down and freezing up in 20-below weather when they're out of reach of a crew. They proved to be an expensive non-solution that, not surprisingly, was abandoned.

The only clue (albeit a substantial one) to 5456's new role as an air repeater car is the blanking of the windshields. Photo by Louis A. Marre.

TEST VEHICLES

During 1979 the Power, Marine & Industrial Division of Electro-Motive equipped former Amtrak SDP40F 569 as "test vehicle" No. 134. Although retaining four of its original six traction motors, the 134 was not used as a locomotive, but had a 12-cylinder 645 fitted for stationary applications. One test, perhaps the principal one for which the locomotive was modified, was a high-altitude trial on the Denver & Rio Grande Western. Number 134 should not be confused with three other SDP40Fs, EMD 169 (ex-Amtrak 609), which in 1983 was used to test the prototype 60-line microprocessor controls, then as the 710 engine test bed; nor with EMD 218 (ex-Amtrak 509), which operated in 1985 and later as a second 710 engine and microprocessor test bed; nor with EMD 268 (ex-Amtrak 531), a 1988 710 engine test bed. Those three are locomotives, though restricted to test operation (as opposed to demonstration). Photo by R. R. Harmen.

SLUGS, MATES, AND FUEL TENDERS

The use of a weighted rail vehicle having traction motors but no power source of its own together with a regular locomotive goes far back into locomotive history. Starting in 1914, the Butte, Anaconda & Pacific used three 40-ton "tractor trucks," single trucks that were coupled to and fed power from BA&P's electric locomotives. Nowadays such units are called slugs.

Because of the limit of adhesion of steel wheel on steel rail, a large diesel locomotive has power in excess of what it can use for tractive effort at low speeds. It is sometimes worthwhile to haul around the weight of a slug to avoid using another locomotive unit. The slug was originally useful only between zero and 12 mph, a range in which the locomotive faces adhesion limitations. In the 1940s old Alco-GE-Ingersoll Rand boxcabs were being cut down to slugs and coupled to diesel switchers for hump service. Slugs had traditionally been used for switching, but the concept of road slugs became practical with the advent of high-horsepower diesels.

Slugs should not be confused with brake trailers (sometimes called sleds), which are similar in appearance. Switching is usually done with the air hoses on the cars uncoupled, which means that all braking must be done by the locomotive. A ballasted brake trailer provides extra wheels tied to the independent braking system of the switching locomotive, spreading the braking effort and reducing the possibility of flattening wheels with a quick stop. Only a few roads have used brake trailers; the common alternative is to use two locomotives in multiple, a six-motor locomotive, or a slug.

YARD SLUGS

Canadian Pacific B100, built by Montreal in 1951, illustrates several features of slug design. A slug must have independent brakes and safety appliances such as handrails, steps, and coupler release levers, and there must be an enclosure or hood above the frame to house traction motor blowers. Because the slug is used in low-speed, maximum-tractive-effort situations, motor cooling is critical, and B100 has numerous air inlets for the traction motor blowers. A slug must be able to sand the rail simultaneously with its locomotive, and if it will be used in any way other than permanently coupled between two units, it must have a headlight. (B100's headlight was apparently salvaged from a steam locomotive.) M. U. connections are provided at both ends of B100; some slugs have the connections at only one end. Photo by Larry Russell.

Chicago & North Western slug BU9 illustrates how all these requirements were met by simply using an old switcher (an Alco S3), de-engined and decabbed. C&NW operated an extensive fleet of yard slugs: BU1-BU9, converted from switchers in 1956 and 1965, and BU10-BU18, converted in 1973 and 1974. Photo by Louis A. Marre.

Length is no indication of whether a slug is intended for road or yard service. Louisville & Nashville 2059 was used solely for yard service, but owed its length to an RS-3 frame. Southern yard slug 2477 was built from scratch, not converted from a locomotive. Both photos by Louis A. Marre.

Southern 2450, 2478, and 2249 illustrate the concept of M. U. through a slug to a second yard unit, in this case a TR2B calf. Current for the slug must come from just one of the two locomotives. Photo by J. David Ingles.

In the rear view of the Richmond, Fredericksburg & Potomac slug, note that there are four traction current cables on the right and two smaller cables on the left. The latter is a connection for the battery-charging circuit, used on many slugs (including the GE MATEs) to run the traction-motor blowers. Photo by Louis A. Marre.

In addition to the usual M. U. connections, bus cables are necessary to carry traction current between the locomotive and the slug, usually two cables for each truck if the slug operates in constant series-parallel or can make transition. The Santa Fe slug pictured above has six motors but needs just four traction current cables. The side view of the connections between a Louisville & Nashville slug and its locomotive shows the usual 27-point M. U. cable, air hoses, and four traction-motor current cables. Locomotives that work with a slug must be specially equipped with these motor connections and are usually referred to as "slug mothers." Both photos by Louis A. Marre.

Canadian National has purchased purpose-built slugs from General Motors Diesel Division. Originally, Class GH-00a and GH-00b (General Motors, Hump, zero h.p.) 260-282 were built in 1978 and 1980 for hump service with GP38-2s, while class GY-00b 451-462 of 1980 were for use with switchers. The GHs had AC blower motors; the GYs had DC. Most of the GYs have been renumbered in the 200-214 series, reclassified GY-00c, and assigned to work with rebuilt GP9s. Photo by Larry Russell.

Santa Fe "drone" (AT&SF terminology for slugs) 3951 and Conrail slug 1102, both cut down from Alco hood units, illustrate the 6-motor yard slug. Such slugs must be powered by a mother unit larger than those usually required for 4-motor slugs. Number 3951 is coupled to 2050 h.p. repowered RSD-15 3900, and 1102 to SD38 6944. Slugs of this size are used mostly in hump yard operations. Both photos by Louis A. Marre.

Some converted road units are hard to tell from powered locomotives except on close examination — the lack of cooling fans and the "S" in the number furnish clues about Union Pacific slug S5, a former GP9B, which is shown paired with SD40 No. 3032. Photo by W. S. Kuba.

Santa Fe continues to build yard slugs from a wide variety of retired units, both four- and six-axle. Yard slug 123 is identifiable at sight as having been converted from a CF7 — its Cleburne-built frame would betray whatever rested atop it. Photo by John B. McCall.

Missouri Pacific's shop at North Little Rock, Arkansas, built 21 yard slugs on SW7 and SW9 frames between 1978 and 1981. Yellow-painted 1407 is an example. Photo by Louis A. Marre.

MATES

General Electric calls its 1971-1972 slug a MATE — Motors to Assist Tractive Effort. GE claims two distinctions between these units and road slugs: First, MATEs operate throughout a broader speed range than typical slugs (up to 30 mph for single-end MATEs, and as fast as the mother units are allowed for double-end MATEs); and second, they are also fuel tenders, preceding Amtrak fuel tender No. 400 by seven years.

MATEs have the same frame dimensions as the U36Bs with which they are paired. They were designed for Seaboard Coast Line's phosphate trains. Pairs of RS-3s had been handling 10,000-ton trains at low speeds. They were replaced by single U36Cs and single-end MATEs 3200-3209, delivered in 1971. In 1972, 15 double-end MATEs, 3210-3224, were delivered to SCL for road service between pairs of U36Bs, where two 6-motor units would otherwise be assigned. The theory was that the U36Bs could take a MATE with them if extra axles were required. They are now CSX 5200-5224.

Seaboard System (now CSX) 5211 is a double-ended MATE. It has headlights at both ends, and each end can be coupled to a locomotive. Photo by Louis A. Marre.

ROAD SLUGS

Chicago & North Western was afflicted with a great deal of track limited to 30 mph or less, good territory for road slugs. In 1970 C&NW created road slugs BU10 and BU11 (later renumbered BU40 and BU41) from Chicago Great Western RS2s, and rebuilt an F3B into road slug BU12 (later renumbered BU30). The slugs were designed to operate between pairs of GP35s, taking power from the controlling unit. C&NW built nine more road slugs, BU31-BU39, from F7Bs in 1971.

Chicago & North Western road slug BU36 was built from an F7B in 1971. Photo by J. Wozniczka, collection of Kenneth M. Ardinger.

Columbia & Cowlitz slug 701B, used with C-415 No. 701, was rebuilt from Burlington Northern Alco S-6 750. Logging railroads are low-speed operations, so the Type A trucks, usually limited to 45 mph, are no handicap. Photo by Kenneth M. Ardinger.

Oregon, California & Eastern slugs 7606 (illustrated) and 7607 were built by Morrison-Knudsen from Union Pacific U25Bs for use with EMD-repowered U25Bs 7601-7605. Photo by Kenneth M. Ardinger.

Southern Pacific's Sacramento Shops built 13 TEBU (Tractive Effort Booster Unit) road slugs numbered 1601-1613 between 1980 and 1982, following the design of Morrison-Knudsen prototype 1600. They were constructed on U25B frames and trucks, have dynamic brakes, and serve as fuel tenders. Photo by Kevin Idarius.

Norfolk & Western constructed more than 30 yard slugs, mostly on 6-motor Train Master frames. Successor Norfolk Southern has created a number of road slugs, utilizing its bottomless supply of retired GP9s and GP18s. Number 9740 illustrates the most recent version, notable for the location of the dynamic brakes. Unlike SP's road slugs, NS's do not have fuel tanks. Since these are single-ended slugs and draw power from only one unit, they are usually deployed in a GP-slug-slug-GP configuration so that both slugs are powered when in use. Road slugs are usually set to cut out at higher speeds than yard slugs, typically in the 30-35 mph range, where the powered units demand all output from their generators. Both photos by George Cockle.

While it may be said that all slugs are unique due to their diverse origins, Kansas City Southern road slugs 4050 and 4055 are "uniquer." KCS has long used de-engined F units as road slugs, but this pair of F3A road slugs is unparalleled in that they are control slugs, not trailers. For use on the daily turn between Shreveport and Cullen, Louisiana, Shreveport Shop in 1982 removed the prime movers from 4050 and 4055, but left the control stands in the cabs. The units operate at either end of a slug-F7-F7-slug set, allowing the engine crew to ride in cool, quiet comfort, at some distance from the laboring 16-567 diesels. One half of the set is shown at Shreveport in 1986. Photo by Louis A. Marre.

In mid-1988 CSX contracted with Precision National for 80 control-cab road slugs to be rebuilt from GP35s and GP30s. The first of these is shown mated with a specially equipped GP40-2. The general appearance of the GP35 is unchanged. The operating cab and the dynamic brakes have been retained, but most hood doors have been welded shut, and inside the hood a block of concrete replaces the prime mover and serves as ballast. These road slugs are also fuel tenders for their "mother" units. The CSX road slugs are significant because the size of the fleet makes it by far the most ambitious application of the concept and because it represents a logical evolution: road slug plus fuel tender plus control cab. Photo by Jay Potter.

FUEL TENDERS

The MATE concept provided a good idea in the form of the fuel tender. Burlington Northern adopted the idea in a big way, placing 65 tenders in service by the end of 1985 after building two prototypes at Northtown (Minneapolis) in 1983. BN operated a large fleet of tank cars to carry diesel fuel from pipeheads to locomotive fueling stations; using them to fuel locomotives directly promised to be less expensive and more efficient. Other potential advantages were elimination of some fueling stations, fuel stops, and the need to bring locomotives back to fueling points from remote assignments — for example, the helpers at Crawford, Nebraska.

The equipment on the tender itself is simple — just hoses and a green locomotive paint scheme. However, the locomotive must be equipped with an extra fuel pump which is triggered by a float in the locomotive's fuel tank. At a pumping rate of 6 gallons per minute, it can fill the tank a little faster than the full load usage rate. Photo by Greg Sommers.

The fuel tenders are designed to feed a locomotive at each end. They must also trainline the M. U. and air connections between the locomotives they serve, and have their airbrake system reconfigured to work with locomotive independent braking. The photo shows these connections. The fuel hose is the nearest one, the one without the air line coupling. Because of this hose connection disconnecting the fuel tender must be done in the shop. Photo by Greg Sommers.

Some of the 1985 fuel tenders were equipped with a saddle tank containing oil for a flange lubrication system, use of which has since been discontinued. Photo by Greg Sommers.

Flange lubrication systems are more commonly mounted directly on the locomotive, such as the extra box behind the blower bulge on this Conrail unit, from which small hoses lead to the wheels, or the small containers positioned just outboard of the brake cylinders on the front truck on Kansas City Southern 706. Flange lubrication is thought to produce a slight saving in fuel, as well as decreasing flange and rail wear, especially on curves. Conrail photo by Greg Sommers; KCS photo by Louis A. Marre.

In 1987 Soo Line placed in service a combined fuel tender and slug numbered 2118, built from an SW1200. The purpose was to spare one of two MP15s usually used on the St. Paul hump and reduce the need to return to the engine house for refueling. Number 2118, which is mated to MP15 1535, may be the prototype of other such units. Soo is also considering road fuel tenders made from locomotive carbodies. Photo by Jerry A. Pinkepank.

MISCELLANEOUS MODIFICATIONS

CABLESS AND HOMEMADE-CAB REBUILDS

As train weights have risen above the 14,000-ton level, especially unit trains, railroads have again considered the idea that not every locomotive in the power consist needs a cab, especially with the rising costs of maintaining control and airbrake equipment and installing FRA-specified glazing. In addition to the purchase of cabless units, notably by Burlington Northern, there has been some rebuilding of locomotives without cabs, usually as the result of wreck damage. Early cases were Missouri Pacific SD40 783 in 1972 and U30C 3311 in 1976, which were temporarily operated without cabs while awaiting parts. However, these units regained cabs and were not prototypes of the trend. The actual start of the phenomenon was the rebuilding of wrecked BN SD40-2 7221 at West Burlington in July 1981 as cabless BN 7500 (briefly numbered 900B). It was soon followed by 2600 (briefly numbered 600B), rebuilt from GP38 2136 (ex-Frisco 660); 2601 (601B), rebuilt from GP38-2 2315; and 4500 (800B), rebuilt from U30C 5336 done in 1981. Three more units followed: 7600, rebuilt from SD40 6302 in 1982; 7501, rebuilt from SD40-2 6914 in 1983, and 7502, rebuilt from SD40-2 6812 in November 1984.

In rebuilding SD40 6302 to SD40B 7600, Burlington Northern tried moving the dynamic brake grids to the former cab location to get them away from the engine heat, thereby easing the cooling task. Photo by R. R. Harmen.

In 1987, BN sold No. 7600 to Soo Line, which numbered it 6450 and used it primarily to run off horsepower hours owed to BN in coal train service. Photo by Jerry A. Pinkepank.

BN 7501 was rebuilt from SD40-2 6914 in 1983 at Livingston, Montana. The old dynamic brake location was retained. Photo by George Cockle.

In 1981 and 1982, BN converted five GP9s to GP9Bs 600-604 to work between pairs of SD7s at the Northtown (Minneapolis) hump yard. Note the paper-air-filter box on No. 601. Photo by Roger Bee.

Santa Fe began a program of rebuilding units without cabs in 1983 at the San Bernardino shops with the creation of SD45B 5501 from SD45 5581. Photo by Greg Sommers.

At the same time Santa Fe created SD45B 5502 from SD45 5523. That unit was wrecked and replaced by second 5502, ex-5340, a previously upgraded SD45. The units were successful. Later units like SD45-2B No. 5513 have had the dynamic brakes relocated to the former cab area, leaving the hood space above the prime mover clear for easier maintenance. Photo by Greg Sommers.

Southern Pacific rebuilt a virtually new B36-7, Cotton Belt 7771, to a cabless unit after it was wrecked in 1982.
Photo by Louis A. Marre.

Missouri Pacific 3319 operated cabless while awaiting parts for a new cab. Photo by J. Harlen Wilson.

Usually a new cab is purchased from the manufacturer or at least rebuilt to the original design, and the result would be indistinguishable from the factory version. However, B30-7A 4607 received a homemade cab that looks more EMD than GE. Photo by George Cockle.

Other roads have occasionally built cabs from scratch, not always to the original design. Maine Central GP7 471 has a homemade version of the crew cab. Photo by Jack Armstrong.

Making a cabless booster unit from a damaged locomotive does not always involve literal removal of the cab. Missouri-Kansas-Texas 401B, a controlless but not cabless F7A, was rebuilt by the road's Parsons, Kansas, shop in 1974 from fire-damaged F7A 66C. The cab windows and number indicators were plated over and a walk-through door was installed in the nose, with a backup light beside it. The unit survived until Union Pacific acquired the Katy in August 1988 and donated the unit to the city of Denison, Texas. Photo by Louis A. Marre.

As a final example of a cabless rebuild, we show one representative of innumerable industrial modifications to full-sized railroad equipment. Inland Steel modified at least four SW1s to cabless remote-controlled units for its East Chicago mill. Inland RC4 is a particularly neat instance of what a well-equipped heavy industrial shop can do. Photo by Louis A. Marre.

HALF PORTIONS

In mid-1980, N&W deactivated half the cylinders of GP9s 903 and 905. The result was an 875 h.p. "GP4½" with the rear radiator set removed. Considerable difficulty ensued from attempts to balance a 16-cylinder crankshaft for operation with only 8 cylinders. The conclusion to the experiment was the Caterpillar-repowering program, covered elsewhere in this book. Both half-Geeps returned to normal configuration by January 1984. Photo by Bob Graham.

RADIAL TRUCK EXPERIMENTS

Since the 1970s there has been discussion in the industry about the merits of using a guided locomotive truck to reduce track stresses. A radial truck, in which the truck frame is articulated and able to guide following axles, is such a design. EMD equipped Santa Fe GP50 3810 with such a truck in December 1984 and Burlington Northern SDP45 6599 in January 1985. Each unit returned to tests on its home road. BN's unit was stored at West Burlington after May 1986, but ATSF 3810 was re-equipped with a second radial design and returned to testing in February 1986. In addition to tracking advantages, such a truck may be able to accommodate a larger traction motor, and speculation has linked the radial truck experiments to A.C. traction motor prospects.

The rear of Burlington Northern SDP45 No. 6599 was equipped with a four-axle radial truck for testing in 1985. Photo by Dan Olah.

THE JOBS LOCOMOTIVES DO

Most of you who are reading this book don't have to make purchasing decisions about locomotives, but if you understand the factors involved in those decisions — factors such as tractive and drawbar force, minimum and maximum speeds, and train resistance — you will better understand the purposes for which the locomotives described in this book were intended and how the locomotives are used.

Understanding why one locomotive has six axles while another has four, why locomotives are ballasted with thick underframes which might seem to be nonproductive weight, and why locomotives which look identical have different train-handling characteristics comes down to understanding simple principles and formulas.

It all begins with train resistance. Even though the steel wheel on the steel rail is the most efficient form of land transportation, friction must still be overcome to move a train from point A to point B. It is the job of the locomotive to overcome that resistance, using the energy which it converts from fuel to driving-wheel rotation.

The friction between steel wheel and steel rail, the enemy to be overcome by the locomotive, is also its ally. Traction requires friction between the locomotive driving wheels and the rail. Friction in this case is not resistance but adhesion.

Train resistance

The most elementary expression of train resistance is that it equals 20 pounds per ton per 1 percent of grade (or a percentage of the weight of the train equal to the percentage of the grade), plus 5 pounds per ton flat rolling resistance, regardless of grade. To express it as a formula:

R = W(G/100 + .0025)
R is train resistance
W is the weight of the train
G is the grade expressed as a percent

(A 1 percent grade rises 1 foot in every 100; a 2.5 percent grade rises 2.5 feet in every 100.)

In practice, the flat rolling resistance factor varies considerably with such things as the temperature of the axle bearings, the condition of the railhead, and the weight of the rail. Rail weight is a factor because lighter rail is depressed more by the wheels, and as a result each wheel is running slightly uphill.

If a 10,000-ton train is to be moved up a sustained 1.2 percent grade, the resistance per ton on the grade is 20 pounds times 1.2 (24 pounds) plus 5 pounds flat rolling resistance — 29 pounds per ton. If 10,000 tons is the trailing tonnage (the weight of the train exclusive of the locomotives) moving the train up the grade will require 29 pounds times 10,000 — 290,000 pounds of drawbar force.

In addition, the weight of the locomotive must be lifted up the grade. Let's say there are three units each weighing 180 tons, 540 tons in all. The same formula applies. It will take another 15,660 pounds of tractive effort to move the locomotive up the grade. Lifting the total weight of the engines and the train up the grade requires 15,660 plus 290,000 pounds of tractive effort: 305,660 pounds.

Tractive effort and drawbar force

Tractive effort is the total force applied to turn the driving wheels; drawbar force is the tractive effort minus the portion needed to move the locomotive.

The most fundamental locomotive power formula states that tractive effort equals horsepower times 308 divided by miles per hour (horsepower is the net horsepower after deducting that needed for cooling fans, air compressor, and such — called parasitic load). If you multiply a rating of 2,000 horsepower by 308 and divide it by 5, 10, 20, 30, and so on, and plot the points on a graph with tractive effort on the vertical axis and speed on the horizontal axis, you will produce a curve that indicates the amount of power the locomotive is delivering to the driving wheels (Fig. 1). At 5 mph the 2,000-h.p. locomotive has a tractive effort output of 123,000 pounds, and at 60 mph it has an output of 10,267 pounds.

Adhesion limit

Figure 1 also shows a similar curve for a 3,000-h.p. locomotive. At

5 mph it has a tractive effort output of 184,000 pounds, and at 60 mph, 15,400 pounds. However, the tractive effort at 5 mph is only theoretical, because the power is high enough to overcome the friction between the driving wheel and the rail, making the wheels slip. Depending on rail conditions and the type of wheel-slip control, a locomotive can deliver no more than 18 to 32 percent of its weight on the driving wheels as usable tractive effort; a traditional rule of thumb for dry, sanded rail is 25 percent. EMD "Super Series" and GE "Sentry" wheel-slip controls allow a factor of adhesion of about 32 percent.

Let's say that our locomotive is a 6-axle machine weighing 180 tons, all of it carried on driving wheels, and that it has conventional wheel-slip control to which the 25 percent adhesion rule applies. The adhesion limit is therefore 25 percent of 360,000 pounds (180 tons) or 90,000 pounds. A horizontal line can be drawn across the graph at the 90,000-pound level, cutting off the speed-tractive effort curve above that point (Fig. 2).

This adhesion limit is one of the reasons 6-motor locomotives are used for heavy hauling jobs instead of 4-motor locomotives. Depending on factors that limit axle load, primarily bridges, most railroads must limit locomotives to a maximum weight of 30 to 35 tons per axle. Therefore, to raise adhesive weight it is necessary to add axles.

If the locomotive in our example were a 4-axle machine weighing 120 tons, all of it carried on the driving wheels, and if the locomotive had conventional wheel-slip control, the adhesion limit would be .25 times 240,000 pounds, 60,000 pounds. The curve for the 4-axle unit would be cut off above 60,000 pounds (Fig. 3).

Note that in either case, the 2,000-h.p. and 3,000-h.p. locomotives have the same theoretical tractive effort at starting, with only the greater adhesive weight making a difference in usable tractive effort. In fact, a 2,000-h.p., 6-axle locomotive on a railroad whose bridges allow 34 tons per axle would have 102,000 pounds of usable starting tractive effort, while a 3,000-h.p. locomotive on a railroad whose bridges allowed only 30 tons per axle would have only 90,000 pounds of usable starting tractive effort, rail conditions being equal. This is why railroads have heavy-haul locomotives ballasted to pro-

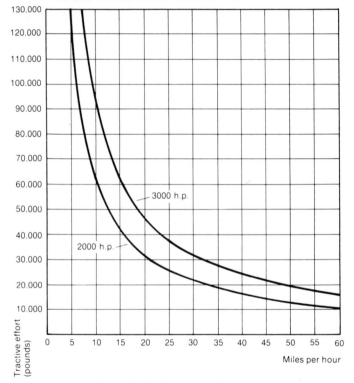

FIG. 1 SPEED-TRACTIVE EFFORT CURVE FOR 2000 H.P. AND 3000 H.P. LOCOMOTIVES

This unorthodox lashup of power on the Kansas City Southern illustrates the leveling effect of low speeds. Sandwiched between the two GP40-2s is a road slug converted from an F7 A unit, and those three units are supplemented by an SW1500. The speed limit on the Leesville-Lake Charles, Louisiana, line where these units are shown, is 45 mph, but speed restrictions of 20 mph and even 10 mph apply in many areas. The 3000 h.p. of the GP40-2s, the 1500 h.p. of the SW1500, and the tractive effort of the slug are all approximately equal in the 10-20 mph range. Each GP40-2 has a continuous rating of 65,000 pounds; the slug, 61,000 pounds; and the SW1500, 62,000 pounds — a total of 253,000 pounds, more than adequate to start the train or to accelerate the train after a speed reduction. However, if the track conditions permitted sustained speeds above 25 mph, neither the slug nor the switcher would be of any use. Photo by Louis A. Marre.

vide the maximum weight on driving wheels if their track and bridges allow it. This ballasting is usually done by fabricating the underframe from thicker steel plates than standard.

Minimum continuous speed

Adhesion is not the only practical limit on the theoretical speed-tractive effort curve. Another is the minimum continuous speed, which is imposed to protect the traction motors from overheating. Locomotive traction motors are series-wound direct-current machines geared to the axle, and their rotational speed is limited by the wheel speed. At any given wattage (and the horsepower of the diesel engine reaches the motors as watts, the product of volts and amps) there is a minimum speed for the motor, below which heat can rise to the point where the insulation breaks down and the motor either shorts out or burns out.

It is mechanically impractical to design a gearshift for the traction motor, so a single gear ratio must suffice. The typical gear ratio is slightly more than 4:1; that is, four rotations of the motor armature for each rotation of the wheels. This is a compromise, because not only must the motor be kept turning above a certain speed, but it must not rotate too rapidly at high wheel speeds, or centrifugal force would tear the motor apart. EMD's usual gear ratio of 62:15 (62 teeth on the wheel or bull gear and 15 on the motor pinion) and GE's equivalent ratio of 74:18 used to allow a top speed of 65 mph, although improved bonding of armatures eventually allowed the limit to be raised to 70 mph.

For a locomotive intended to operate over the normal range of

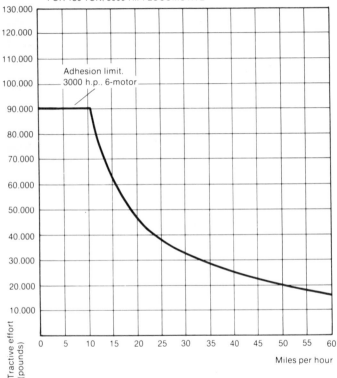

FIG. 2 SPEED-TRACTIVE EFFORT CURVE
FOR 180 TON, 3000 H.P. LOCOMOTIVE

Adhesion limit,
3000 h.p., 6-motor

Tractive effort
(pounds)

Miles per hour

freight-train speeds, the combination of the maximum speed and the gear ratio determines the minimum allowable continuous speed. That term implies the further phrase, "at full rated horsepower." If train tonnage pulls the speed down, the result of throttling back to avoid traction motor heating will be that the train stalls. Here the 6-motor locomotive has another advantage: The wattage produced by the generator is spread over 6 traction motors rather than 4, allowing a lower continuous speed for the same horsepower and gear ratio.

Typically a 4-motor, 3,000-h.p. locomotive with a 62:15 gear ratio has a minimum continuous speed of 13 mph, while a 6-motor, 3,000-h.p., 62:15 locomotive has a minimum continuous speed of 9 mph. Because the speed-tractive effort curve rises steeply between 9 and 13 mph, the combination of the lower minimum continuous speed and the higher permissible weight on drivers lets the 6-motor locomotive reach its adhesion potential at around 10 mph with 90,000 pounds of tractive force. A 4-motor locomotive, even using the newest wheel-slip controls, would be limited to just over 70,000 pounds of tractive effort at 13 mph to keep the motors from overheating (Fig. 4).

This is an oversimplified statement of what happens with the new wheel-slip controls, since they achieve their high effective adhesion by rapid fluctuations in power output, and in effect vary the horsepower each microsecond to meet the adhesion conditions, seeking at any given instant the highest power input that the adhesion conditions will stand. This has been described as allowing the wheels to creep rather than slip.

In addition, the newest locomotives use power-limit features to prevent motor overheating. The output to the traction motors is lowered to stay within heating limits. A locomotive operating under the power-limit feature is likely to be operating below the adhesive limit as well, and a new factor in the equation with modern power is to be sure the power-limit feature isn't actually keeping the locomotive from doing its job.

Which locomotive? How many?

Let's examine an imaginary railroad for which a number of locomotives are to be purchased. This railroad includes a river-grade line over which freight and passenger trains move at high speed, a light-density branch with several weak bridges, and a heavy-duty branch that has a 20-mile-long, 1.5-percent uncompensated grade with successive 10-degree curves. (We'll explain the engineering terms as we discuss the factors they affect.) The grade descends to a coal mine, so loaded trains must climb it. With these considerations in mind we can approach the question of purchasing appropriate locomotives.

Our initial objective is to avoid specialization. In the first wave of dieselization in the 1940s and 1950s, railroads often purchased specialized locomotives for particular tasks. They found that many of these units weren't used much and later required modification for the railroad to get full return from its investment. Then railroads turned to hood-type locomotives in the 1,500- to 1,800-h.p. range as building blocks for all types of service, and resorted to other measures such as standardizing on a single type of multiple-unit control or placing road trucks and multiple-unit control on switchers.

When higher-horsepower locomotives and 6-motor mainline power became common in the 1960s and 1970s, specialization returned to a degree, but it was overwhelmed by the use of 6-motor, 3,000-h.p. locomotives for a wide range of jobs. The SD40 and the U30C assumed the same building-block role as the first generation's GP9 and RS-3. This led to the purchase of over 4,000 SD40s and SD40-2s and over 1,300 U30Cs and C30-7s in the U. S. alone, so that by 1987 well over one-fourth of the road locomotives in the U. S. were 3,000-h.p., 6-motor machines. However, the jobs requiring such locomotives were considerably less than one-fourth of the total. In many assignments the 6-motor configuration added nothing to performance; indeed, it was a drawback because of the fuel needed to move the extra weight and the maintenance cost for the extra traction motors.

To return to our hypothetical railroad: If we buy our locomotives based on the thinking that prevailed in the 1970s, we may well purchase 6-motor locomotives in the 3,000- to 3,900-h.p. range, because they can run on our coal branch with its 20-mile, 1.5 percent grade and 10-degree curves. However, we will have to keep their axle load down if we want to use those same units on the light-density branch with its limiting bridges — and that will reduce coal-hauling capacity.

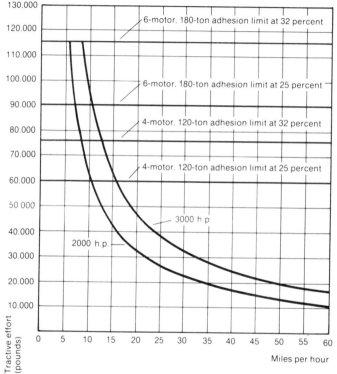

FIG. 3   SPEED-TRACTIVE EFFORT CURVES FOR 120 AND 180 TON, 2000 H.P. AND 3000 H.P. LOCOMOTIVES

Assigning helpers to assist trains over short ruling grades can substantially reduce locomotive miles. A good example is the Kansas City Southern main line between Kansas City, Missouri, and Port Arthur, Texas. Most of the line has moderate grades, but 30 miles of the Second Subdivision between Dalby, Missouri, and Gentry, Arkansas, have 1.5 percent grades in each direction and numerous curves. There is a similar stretch of line on the Fourth Subdivision between Page and Rich Mountain, Oklahoma. Over most of the Kansas City-Port Arthur route five units are sufficient for 14,000-ton coal trains, but seven units are necessary for those trains on the grades. Rather than invest in remote-controlled midtrain slave units or put seven units at the head of the trains (which presses the limit of drawbar strength), Kansas City Southern assigns a pair of 6-motor pushers for the grades. The photos, taken at Page, Oklahoma, show five 6-motor units — three KCS SD40-2s and two Burlington Northern GE units — at the head of a train, and two SD50s pushing at the rear. Photos by Louis A. Marre.

A second objective is fuel and maintenance economy, and this conflicts with our desire to buy a single universal locomotive. A 6-motor locomotive will obviously burn more fuel to carry around 60 or more extra tons per unit on jobs where the extra weight is unproductive because the additional adhesive weight is unnecessary and a low continuous speed is not expected. The two extra traction motors and wheel sets also increase the annual maintenance cost of the locomotive by at least 10 percent.

If we follow an emerging pattern of thought in our motive power purchase, we will look for a locomotive which can run singly on the light-density branch line, be run in multiple to meet the speed and power requirements of the river-grade main line, and still have enough drawbar force to function on the coal branch without adding excess units. To do all this, we will have to take full advantage of the higher adhesion made possible by the new wheel-slip controls.

Different trains — different locomotives?

Let's look at the standard trains on our railroad. The coal train will consist of 110 cars (no caboose), each weighing 131 tons loaded, for 14,410 trailing tons. The speed limit for this train on the main line is 45 mph, and the otherwise-level main line has a 5-mile, .7 percent compensated grade that loaded coal trains will have to climb. (We've already mentioned the 20-mile grade on the branch.) The standard merchandise train is a heavyweight: 120 cars at an average of 65 tons each, 7,800 tons in all. We limit this class of train to 50 mph. Our third standard train is a TOFC (Trailer On Flat Car) train. TOFC trains are limited to 3,000 tons, and their maximum speed is 60 mph. On the light-density branch line, bridges limit the weight per axle to 28 tons. The maximum train to be handled on that branch is 3,000 tons at 25 mph, and the maximum grade is .5 percent, compensated, sustained for 2 miles.

For reasons of fuel economy and to reduce maintenance expense, we choose a 12-cylinder, 4-motor locomotive with a modern wheel-slip control system, such as the 3,000-h.p. General Motors GP59 or the 3,200-h.p. General Electric B32-8. We'll plot the trains on our speed-tractive effort curve to see how many such locomotives are required for each.

The Davis Formula

Our methodology for calculating train resistance is the Davis Formula, named for W. J. Davis of General Electric, in whose article, "The Tractive Resistance of Electric Locomotives and Cars" (*General Electric Review*, October 1926), the formula first appeared. It was a refinement of similar work done previously, particularly by A. M. Wellington in his 1887 book, *Economic Theory of the Location of Railways*. Although much work has been done to refine the Davis formula, it remains the fundamental tool in train-resistance calculation. The formula is:

$R = 1.3 + 29/W + 0.045V + 0.0005AV^2/WN$
R is resistance in pounds per ton
A is the cross-sectional area of the average car in square feet
W is the weight in tons per axle
V is the speed in miles per hour
N is the number of axles per truck

The term **$1.3 + 29/W$** is the resistance independent of speed, a refined version of the flat rolling resistance of 5 pounds per ton given at the beginning of this appendix. There is a constant resistance of 1.3 pounds per ton plus a variable equal to 29 divided by the weight in tons per axle. The factors of 1.3 and 29 are empirical constants, derived from field testing. This part of the formula says that light cars have a greater relative rolling resistance — resistance *per ton* — than heavy cars. It makes sense if you consider that flat rolling resistance is a matter of friction and is largely independent of the weight on the wheel. It is also one of the reasons that modern 100-ton-capacity cars are more efficient than the 40- and 50-ton-capacity cars of years past.

The second part of the formula, **$0.045V$**, is the portion of train resistance that is related to the first power of the train speed. Graphed, it is a straight line angling upward from the origin. At 5 mph it is .225 pounds per ton; at 60 mph it is 2.7 pounds per ton.

The third part of the formula, **$0.0005AV^2/WN$**, is the effect of wind resistance as felt at the wheel. It reflects the fact that train resistance increases at a greater rate than the speed. It takes 2.4 horsepower per ton to move at 60 mph on level track the same train that

FIG. 4 SPEED-TRACTIVE EFFORT CURVES
FOR 120 AND 180 TON, 2000 H.P. AND 3000 H.P. LOCOMOTIVES

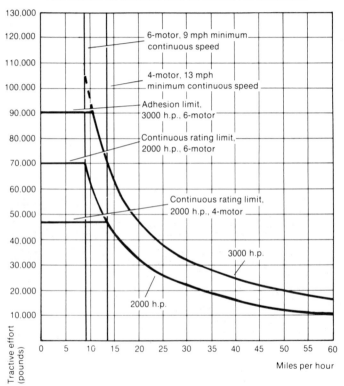

1.6 horsepower will handle at 50 mph — a 20 percent increase in speed requires a 50 percent increase in horsepower.

Applying the formula

Applying this formula to the typical trains described above yields the three curves of Fig. 5. The train with the highest resistance is the coal train. If we assume a cross-sectional area of 100 square feet, the train resistance will be 39,195 pounds at 10 mph, 48,994 pounds at 20 mph, 60,954 pounds at 30 mph, 75,076 pounds at 40 mph, and 82,948 pounds at 45 mph. To achieve track speed of 45 mph on level track, this train will need power capable of exerting nearly 83,000 pounds of drawbar force. A 3,000-h.p. locomotive, regardless of the number of axles, is exerting about 20,000 pounds of tractive effort at that speed, not counting what is needed to move the locomotives themselves. If we assign four GP59s, the coal train will be able to hit 40 to 45 mph on level track.

If we apply the formula to the typical 7,800-ton merchandise train, assuming an average cross-sectional area of 120 square feet, we find that the train has a resistance of 28,938 pounds at 10 mph, 36,738 pounds at 20 mph, 47,502 pounds at 30 mph, 61,074 pounds at 40 mph, and 77,454 pounds at 50 mph. To achieve 50 mph we will need five 3,000-h.p. units, or we can assign four units and accept a speed a little above 45 mph — a little faster downhill. The latter is what railroads do most often. Except for the fastest schedules, trains are powered so they can achieve track speed only on the downhill portions of the line.

Finally, for the piggyback (TOFC) train, we have to nearly double the calculated cross-sectional area per car to represent the trailers (2 trailers per car, 30 cars with 60 trailers), so we assume a cross-sectional area of 180 square feet and an average weight of 105 tons per car — 3,150 tons, slightly over our limit of 3,000 tons. The resistance of the train is 9,482 pounds at 10 mph, 12,537 pounds at 20 mph, 16,632 pounds at 30 mph, 21,830 pounds at 40 mph, 28,130 pounds at 50 mph, and 35,469 pounds at 60 mph. It will take three 3,000-h.p. units to move this train at 60 mph on level track.

Horsepower per ton

Another way to state the results of the resistance calculations is to

If moving a train over the road economically is the sole objective and speed is not important, even a single unit can perform satisfactorily as long as it does not drop below the minimum continuous speed. On the flat Toledo Division of the Baltimore & Ohio (now CSX) a single GP40-2 can move a 100-car train of mixed freight at about 20 mph. Two more units would be required to achieve a sustained speed of 50 mph, but the 65,000-pound tractive effort of the lone GP40-2 is enough to start the train and keep it moving. Photo by Louis A. Marre.

translate them to horsepower per ton. Divide the train resistance at a given speed by the drawbar force for that speed for the class of engines being considered to determine the number of engines required; multiply that by the horsepower rating; and divide by the number of tons.

For example, our 7,800-ton merchandise train requires 77,454 pounds of drawbar force at 50 mph. Using the Davis formula, we can calculate that a 3,000-h.p. GP59 weighing 120 tons and having a cross-sectional area of 150 square feet requires 2,042 pounds of tractive effort to move itself at 50 mph. Its calculated tractive effort of 18,480 pounds at 50 mph becomes a drawbar force of 16,438 pounds. Dividing the 77,454 pounds of train resistance by the 16,438 pounds of drawbar force gives a figure of 4.71 locomotive units. Multiplying that by 3,000 h.p. and dividing by 7,800 trailing tons yields 1.81 horsepower per ton.

It is possible to treat the locomotive weight as part of the train weight rather than do a separate calculation for drawbar force. Such a calculation yields requirements of 1.02 h.p./ton at 40 mph, 1.6 h.p./ton at 50 mph, and 2.4 h.p./ton at 60 mph. By contrast, the heavily loaded coal train requires only .68 h.p./ton at 40 mph because it has a high concentration of weight in relation to its cross-sectional area. The TOFC train of our example is also heavily loaded and can get along with 2.19 h.p./ton at 60 mph. Lightly loaded TOFC trains can often require more than 4 horsepower per ton to move a train of empty trailers at 60 mph.

Graphing the train resistance curves shows that the reason for several-unit locomotive consists is speed, not hauling power. One 3,000-h.p. locomotive could handle our 14,410-ton coal train at about 18 mph, if it could start the train.

Starting resistance

Starting resistance of trains is something of a mystery, despite years of experience. The Davis formula doesn't go down to zero miles per hour. Starting resistance of cars has been estimated at anywhere from 14 to 54 pounds per ton; the usual figures are in the 25- to 35-pound range. The figure depends on the temperature of the axle bearings, how far the cars are depressing the railhead, the amount of skewing of the trucks (which increases flange resistance), and many other factors.

The figures at the high end of the range for starting resistance lead to a question: Could trains start at all if slack were eliminated? If the 54-pound figure were correct, a 6,000-ton train would require 324,000 pounds of tractive force to start on level track, almost enough to threaten breaking the lead drawbar, the nominal capacity of which is 360,000 pounds for Grade C material. (Unit trains are often made up entirely of cars having Grade E couplers and draft gear, which have a designed capacity of 500,000 pounds.)

In practice, trains do not start all at once because of the cushioning provided by the draft gears and slack between the couplers. The train is usually stopped with the slack run in. If not, an engineer will take slack by backing into the train. You may see an engineer attempt to start a train which has had the slack stretched out of it by a slight grade. The engineer will set the brakes, back against the train to get some slack, then try to time the start so the release of the air brakes coincides with the arrival of the pulling power of the engine at each drawbar. This is at best an exercise in guesswork — a stall against unreleased brakes at the rear of the train or a break-in-two will be the result of guessing wrong.

For purposes of assigning power it is usually assumed that trains will be able to take slack. In any case, power that has been assigned for speed can start any train it can move at track speed. This is the opposite of the drag-freight era steam power axiom, which stated that a locomotive could move any train that it could start. When railroads began paying attention to locomotive horsepower in the 1920s and freight-train speeds began to increase, starting tractive effort ceased to be the main criterion in locomotive design.

Power for the grade

Now that we have examined the power for our railroad from the standpoint of speed on level track, we can move to the question of grade. Can the assigned power handle the train on the ruling grade without stalling and without pulling out drawbars?

There is also a related question of whether the train can be started on the ruling grade. The answer does not necessarily determine the

Computer simulation is a convenient tool for estimating locomotive performance, but road testing provides empirical verification of the results. EMD and GE have test cars, and a few railroads still have dynamometer cars (usually leftovers from the age of steam, updated to measure diesel performance). In May 1987 Soo Line tested EMD GP59 demonstrators 8, 9, and 10, and GE B32-8s 5497, 5498, and 5499 (painted and lettered for Burlington Northern, but GE demonstrators nonetheless). The two sets are shown coupled together ahead of a Soo Line SD40-2, Soo's ex-Milwaukee Road dynamometer car X5000, and a carefully weighed train. The competing sets of locomotives were tested individually at selected locations. Photo by Roger Bee.

FIG. 5 TRAIN RESISTANCE CURVES

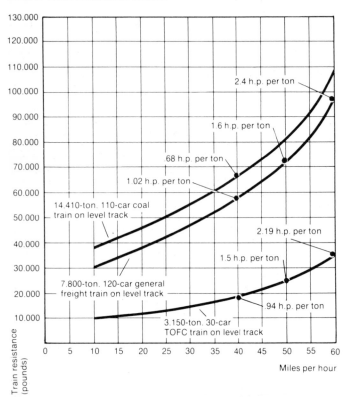

Train resistance (pounds) vs. Miles per hour

2.4 h.p. per ton
1.6 h.p. per ton
.68 h.p. per ton
1.02 h.p. per ton
14,410-ton, 110-car coal train on level track
2.19 h.p. per ton
1.5 h.p. per ton
7,800-ton, 120-car general freight train on level track
.94 h.p. per ton
3,150-ton, 30-car TOFC train on level track

powering decision. If stalling on the grade is deemed an infrequent occurrence, it may be practical to let the train double out of difficulty — cut the train in half, take the front part to the top of the grade, then go back for the rear part.

The ruling grade on our main line is 5 miles of .7 percent. The coal train's 14,410 trailing tons have a resistance on the grade of 19 pounds per ton (.7 times 20 is 14 pounds for the grade, plus 5 pounds per ton constant rolling resistance) for a total resistance of 273,790 pounds. At a minimum continuous speed of 13 mph, four GP59s produce a tractive effort of 284,000 pounds. They require 9,120 pounds to move themselves up the grade, leaving 274,880 pounds of drawbar force. Therefore, the four units will be just sufficient to take the coal train up the grade without a helper and without falling much below the minimum continuous speed.

Short-time ratings

As the grade is short, we can probably tolerate being at 12.5 mph in the short-time rating of the locomotives. This is known as running in the red because the dial of the locomotive ammeter is red above a certain point. There is a range of amps that is permissible for an hour, another for 45 minutes, another for 30 minutes, another for 15, another for 5. Wise power planning uses these short-time ratings judiciously. (On the newest locomotives the engineer is not expected to regulate this; the locomotive goes into the power-limit mode automatically.)

If we had older 3,000-h.p., 4-motor locomotives without the newest wheel-slip control, the locomotives would be adhesion-limited to 60,000 pounds each. Four units would be limited to 240,000 pounds. The coal train would require a fifth unit (perhaps a helper just for the grade) or 6-motor locomotives with a minimum continuous speed of 9 mph. On days with poor adhesion conditions or increased train resistance (high winds, for instance) the train would stall with four units. In any case, the amount of power required does not threaten the 360,000-pound working limit of the Grade E steel couplers (500,000-pound nominal rating) on the unit coal train. The train will not break apart because of the power applied at the leading drawbar.

However, if the train had to be started while stretched on the grade, the starting resistance would be 565,000 pounds (39 pounds per ton — 25 pounds constant and 14 pounds for the grade). The train would have to be doubled off the hill, even with a fifth unit, because the force required to move the train would exceed the strength of the couplers.

The 7,800-ton merchandise train has a resistance of 148,200 pounds on the ruling grade. Four units are ample, and if we examine the speed-tractive effort curve, we see that this grade will pull down the speed of the train until the power balances the train resistance at 22 or 23 mph. This can be determined by adding the resistance of the locomotives themselves on the grade, 9,120 pounds, for a total resistance of 157,230 pounds. Dividing that figure by 4 yields 39,330 pound of tractive effort required per unit. An examination of the speed-tractive effort curve for the 3,000-h.p. locomotive shows that this power is achieved at 22 to 23 mph. If the train is below that speed on the grade, it will accelerate to that speed; if it is going faster when it hits the grade, it will slow down to that speed.

There is no threat to the drawbars even with a working limit of 260,000 pounds for the 360,000-pound rating of the Grade C steel couplers of this train. At 39 pounds per ton, the starting drawbar force for a stretched train on this hill would be about 304,200 pounds, well within the designed rating of the Grade C steel, although above the shock-load rating of 260,000 pounds. We can assume that the stretched train could be started on the hill without doubling, since shock loads will not occur in this situation. (There will be acceleration loads, which are discussed later, but train resistance will have dropped from its starting peak before acceleration resistance becomes a factor.)

Four different conclusions!

Given the analysis above, four different railroad transportation departments could reach four different conclusions. An economy-minded management, to whom the last mile-per-hour of speed and the occasional delay for doubling the grade are less important than reduced expenses, will choose 4-unit sets of GP59s or B32-8s. This choice reduces both initial investment and maintenance expenses and allows the lowest transportation cost, assuming crews can still make it over the road within the Hours of Service Law or without incurring excess overtime.

A management that emphasizes meeting schedules despite higher costs might choose to add a fifth unit to the coal and merchandise trains to be able to attain track speed more of the time (four units were just shy of being able to make 45 mph with the coal train and 50 mph with the merchandise train on level track). The fifth unit will also help avoid the risk of delay to the coal train (and to other trains on the line at the same time) under marginal adhesion conditions.

A third management with a cautious approach to the coal train but otherwise prepared to be economical might choose to forgo standardization and buy 6-motor locomotives for the coal train and stay with four GP59s on the merchandise train. Yet a fourth management, relying on typical 1970s thinking, would purchase 6-motor, 3,000-h.p. locomotives for both assignments in order to preserve standardization, even though the extra traction motors are no help to the merchandise train.

Heavy grades and weak bridges

Now let's examine the 20-mile, 1.5 percent grade on the branch to the coal mine. The grade has a succession of uncompensated 10-degree curves. "Uncompensated" means that the rise of the roadbed has not been abated to make up for the additional resistance produced by the curve. That resistance is equivalent to a grade of .04 percent per degree of curvature. (The degree of curvature is the angle through which the track turns in 100 feet. To measure a curve without surveyor's instruments, stretch a 62-foot long string from point to point on the gauge side — the inside — of the outer rail. The distance from that rail to the midpoint of the string — 31 feet from each end — in inches equals the number of degrees of the curve.) Multiplying the .04 by 10 gives another .4 percent to make the grade an effective 1.9 percent.

On this grade a 14,410-ton train has a resistance of 619,630 tons (1.9 times 20 is 38 pounds per ton grade resistance plus the constant 5 pounds per ton; 43 pounds per ton in all). Coal trains can't be brought up the grade intact with locomotives only on the head end, because the couplers are rated for only 500,000 pounds, and we must allow some margin below that for shock loads. It will be necessary to double the hill or to use pushers. On the grade a 4-motor GP59 has a drawbar force of 66,396 pounds at 13 mph, assuming 27.7 percent adhesion with the Super Series wheel-slip control. The train therefore will need ten such units, five pulling and five pushing. If 6-motor, 3,000-h.p. units are used instead, with a drawbar force of 96,367 pounds at 9 mph, seven units will be necessary, four pulling and three pushing.

Unless the coal trains were frequent, a railroad would be unlikely to purchase five new GP59s for pusher duty. The railroad determined to economize would probably choose to double the hill, although the crew time and the lap-back miles would have to be weighed against the cost of providing the helper. The railroad which chose to split its fleet and provide 6-motor units for the coal train would be in the best position, as it could station a three-unit helper, probably older 6-motor, 3,000-h.p. units, on the mine branch. The 1970s-style management which used such units on everything would be in good shape, except for the costs of maintenance and fleet ownership.

An option that is not available is to use 6-motor helpers at 9 mph to assist 13-mph-minimum road engines. The result would be either burned-out traction motors on the road engines from lugging up the grade at 9 mph or running the 6-motor engines at 13 mph, where there is no advantage in having the extra two motors.

On the light-density branch line, a single GP59 can handle a 3,000-ton train at 25 mph, except that on the 2-mile, .5 percent grade the speed will drop to 20 mph. At no point is the adhesion limit pushed. Many managements would nevertheless use two such locomotives to avoid turning them at the end of the line, as failure insurance, and to provide more rapid acceleration after local stops to hold down crew overtime and avoid Hours of Service problems. However, such practices, common in the past, will be questioned as locomotives become more expensive, more powerful, and more reliable. The number of trains being handled by single new, powerful units will probably increase.

Our light-density branch has weak bridges. Weight for adhesion is important for the coal and merchandise trains, and were we to specify lighter-weight units for our entire fleet, we would need more units for those trains. It makes sense to retain older, lighter locomotives to serve the branch, buy a specialized locomotive for this line, or strengthen the bridges so specialized locomotives aren't necessary.

The acceleration factor

The TOFC train can reach its track speed with three GP59s or B32-8s, and the only question might be that of acceleration. Let's say that we want our TOFC train go from 0 to 60 in 5 minutes. The formula for calculating the acceleration force required is:

$Fc = 2000W/g \times 1.05 \times a$
Fc is the correction force in pounds
W is the weight of the train in tons
g is 32.16 (the accelerative force of gravity)
a is the acceleration in feet per second per second

The acceleration factor **a** is calculated by taking the difference between speeds at the beginning and the end of the interval being measured and dividing it by the number of seconds in the interval. In our case, the difference is 88 feet per second (60 mph equals 88 feet per second) and the interval is 300 seconds, so **a** equals .2933. For our 3,150-ton TOFC train, the formula applies a correction force of 60,238 pounds to the zero force existing at zero mph in order to achieve 60 mph in 5 minutes on level track. Once it reaches 60 mph, it will require only 35,469 pounds to maintain that speed.

However, our trio of GP59s can't maintain that rate of acceleration all the way to 60 mph, because by 55 mph the three are putting out only 50,400 pounds of tractive force. It will take a fourth unit to sustain this rate of acceleration to 60 mph.

How much power does it take to accelerate from 55 mph to 60 mph in one minute? The difference is from 80.67 feet per second to 88 feet per second in 60 seconds, .122 feet per second per second. The formula shows the correction force to be 25,135 pounds, which must be added to the force necessary to sustain 55 mph, 31,689 pounds, giving a total of 56,824 pounds — and such power is available only from four units.

The acceleration factor causes schedule-conscious roads to add more power than is necessary to sustain track speed in order to regain speed quickly after meets, cautionary signals, slow orders, and the like.

Low speeds

At the other end of the speedometer the adhesion limit is especially vexing. It is here that slugs have found their application. Slugs are engineless locomotive frames with traction motors on their axles and power cables connecting them to a mother unit. Consider, for example, a 12,000-ton train moving on a branch line where the speed limit is 10 mph because of track conditions and the ruling grade is .5 percent. On that grade the train will have a resistance of 180,000 pounds. Two 3,000-h.p., 4-motor locomotives can generate a tractive effort of 184,100 pounds at 10 mph. Deduct 3,600 pounds to move the locomotives, and the drawbar force is 181,200 pounds. But the units cannot work continuously at full power at 10 mph, and in any event they are limited by adhesion to 120,000 pounds of tractive effort.

By adding a slug, the watts of output are spread over 12 axles, rather than 8, allowing a continuous speed of 9 mph. If the slug and the mothers weigh 120 tons each and can sustain 25 percent adhesion on dry, sanded rail, the adhesion limit is raised to 180,000 pounds, enough for the train.

The slug is most useful at very low speeds, which is why they are usually found in yards rather than out on the road. At 25 mph, about the highest speed at which slugs are of any value, a 3,000-h.p., 4-motor locomotive has a tractive effort of 36,960 pounds, which is so far below the 60,000-pound adhesion limit that the adhesion would have to drop to 15 percent before a slug would be able to add any drawbar force. The only real use for road slugs is on grades that might otherwise pull trains below the minimum continuous speed.

Computers

Who sits down and does all these calculations on the railroad? In steam days, the task was performed by the mechanical and engineering departments, but for more than 20 years these formulas have been embedded in computer programs known as Train Performance Calculators. The programs are coded with the characteristics of all the main lines and significant branches of a railroad. When the railroad needs to change its operations or acquire locomotives, the standard trains and the characteristics of the locomotives under consideration are coded, and the computer simulates the operation. The runs are studied for instances of stalling, dropping below the minimum continuous speed, and excessive drawbar force, and for running times. The results help the railroad make intelligent decisions about motive power.

GLOSSARY

ASEA: Allmänna Svenska Elektriska Aktiebolaget (Universal Swedish Electric Co., Ltd.), Swedish locomotive manufacturer, recently merged with the Swiss firm Brown Boveri to form ASEA Brown Boveri.

Availability: the time a locomotive is free from disabling mechanical defects between scheduled maintenance dates. It is usually expressed as a percentage: "Engine 4900 was 95 percent available in March" means mechanical problems kept it out of service 1.5 days.

Ballast: any extra weight added to a locomotive to bring it up to a desired weight. An SD40-2, for example, can weigh anywhere from 150 to 200 tons, according to the buyer's specification. The usual method of ballasting is to use thicker sheets of steel to fabricate the frame. Occasionally smaller amounts of weight, usually concrete castings, are added to equalize weight distribution.

Blower duct: EMD GP and SD units built since the GP30 have a rectangular duct leading from the central equipment blower, just below the central air intake, straight down to the running board on the left side of the unit (short hood considered front). The shape was changed to slope outward from top to bottom during late Dash Two production. A horizontal duct continues to the rear atop the left running board. The ducts carry air to cool the traction motors.

Bore: cylinder diameter.

Canadian cab or

Comfort cab: any control cab that is significantly larger than what has been used for the past several decades and which incorporates any of several features creating a better work environment for the engine crew, such as acoustic and thermal insulation, air conditioning, and improved collision protection.

Continuous rating: maximum amperage at which traction motors may operate continuously without overheating. It is the effective electrical limit on locomotive performance. See short-time rating.

Contract rebuilding: locomotive rebuilding performed by an outside firm rather than by the railroad's own shop. See page 240.

Dash 2: convenient nomenclature for referring to EMD models produced from 1972 to 1987: 645 engine, modular electronics, and, later, improved wheel-slip control, but without microprocessors.

Dash 7: convenient nomenclature for referring to GE models produced from 1976 to 1988: numerous upgraded components, but no radical departures from GE's U line of 1961.

Dash 8: convenient nomenclature for referring to GE models produced in 1984 and later: microprocessor to control engine and electrical functions, new hood and cab design, improved wheel-slip control system.

Displacement: the volume displaced by a complete stroke of the piston. It is the product of the cylinder area and the piston stroke.

Drawbar force: total tractive effort minus the tractive effort needed to move the locomotive.

Dynamic braking: originally called regenerative braking when applied to electric locomotives, it is a system for temporarily employing traction motors as generators and using the resulting electromotive force to retard the train. On some electric railroads the current produced was returned to the power distribution system. On diesels the current is passed through resistors which convert the energy to heat, which is dissipated into the air by fans.

Extended-range dynamic braking: an optional high-amperage system that allows dynamic braking down to lower speeds than the standard system (below 15 mph).

Factor of adhesion: ratio of locomotive weight to tractive force. A 400,000-pound locomotive exerting 100,000 pounds of tractive force has a factor of adhesion of 4.0.

HEP or

Head-end power: electric power supplied from the prime mover of the locomotive, a special generator on the locomotive, or a car equipped with generators for heating, lighting, and air conditioning passenger cars.

High-adhesion truck: any locomotive truck designed to improve wheel-to-rail contact, usually by minimizing weight transfer between axles caused by varying combinations of forces acting on the truck.

Horsepower hour: When locomotives began to be leased on a performance-use basis rather than on time alone, a new measurement was devised, the horsepower hour, recorded by on-board microprocessors. An LMX Dash 8-39B (see page 270) operated for an hour at full power runs up a charge of 3900 horsepower hours.

Hours of Service Law: a federal regulation specifying a maximum of 12 hours continuous duty for train crews.

Lease turnback: many locomotives are obtained by railroads through lease agreements with financial institutions or leasing companies. When these expire, usually after 15 or 20 years, the railroad may buy the unit, renew the lease, or return it to the lessor — the last is called turnback. Massive lease turnbacks became a feature of locomotive fleet management in the 1980s.

Liberated exhaust: on nonturbocharged EMD locomotives, retrofitted exhaust stacks in excess of the two as built — usually four.

Microprocessor control: miniaturized computers on board the locomotive monitoring and directing the prime mover and the electrical apparatus.

Modular electronics: any "black box" system whereby the electrical control apparatus of the locomotive is subdivided into plug-in components that can be replaced without disturbing other components.

Nominal: advertised. When being tested on the road, a 3000 h.p. locomotive may produce only 2800 h.p. because of mechanical problems, climate extremes, or poor fuel, but its nominal rating is still 3000 h.p.

Parasitic load: any of the loads or devices powered by the prime mover that do not contribute to tractive effort, such as air compressor, traction motor blower, and radiator fans.

Performance curve: see Jobs Locomotives Do, page 315.

Power assembly: the related parts of each cylinder of a diesel engine that can be replaced or repaired without disassembling or removing the entire engine: cylinder head, cylinder liner, intake and exhaust valves, fuel injectors, etc.

Prime mover: the main power plant in an internal combustion locomotive, as distinct from auxiliary power plants for head-end power, air compressor, and such. The term is useful because "engine" is also a synonym for locomotive and can be confusing in some contexts.

RCE: Remote Control Equipment.

Retired: a general term that means off the roster; no longer on the railroad's list of locomotives, farther out of service than "stored" but possibly not yet scrapped.

Ruling grade: the steepest or longest grade on a given section of railroad, dictating or ruling the amount of tractive effort required to move a given tonnage over that line. See tonnage rating, below.

Short-time rating: maximum allowable amperage for a locomotive over a limited period of time — 15-minute rating, 30-minute rating, etc. — beyond which the traction motors will be damaged. It is higher than the continuous rating, defined above.

Slug: a weighted rail vehicle having traction motors but no power source. It is powered by electricity drawn from a regular locomotive. A diesel-electric locomotive produces more electric power than its traction motors can absorb at low speeds; a slug produces tractive effort from that excess power.

Steam generator: a compact boiler installed in a diesel or electric locomotive to provide steam for heating passenger cars. It burns diesel fuel from the main fuel tank. It requires a substantial supply of water, since typical consumption is 1500 to 3000 pounds (200 to 400 gallons) per hour.

Stroke: the distance a piston moves in a cylinder.

Tonnage rating: the maximum permissible trailing load for a given unit or locomotive over a given piece of line. It depends on tractive effort and ruling grade.

Traction motor: an axle-mounted electric motor which provides tractive power to the wheels on electric and diesel-electric locomotives. Conventional traction motors are DC; AC traction motors are being tested.

Tractive effort: the total force applied to turn the driving wheels of a locomotive.

Unit reduction: the principle of reducing the number of units required for a given job by increasing horsepower per unit or effective adhesion; for example, replacing three 1500 h.p. F3s with two 2250 h.p. GP30s or replacing three SD40-2s with two SD60s. The SD60 has 800 more horsepower than the SD40-2 and a sophisticated wheel-slip control system.

Wheel-slip control system: See Jobs Locomotives Do, page 315.

Zombie: a term used by motive power enthusiasts to denote a locomotive whose frame and trucks have been used as the basis for a low-power re-engining.

567: designation of the engine used by Electro-Motive from 1938 to 1966. Its 8½″ bore and 10″ stroke give it a displacement of 567 cubic inches per cylinder. (Area = πr^2. Radius = 4.25″; the square of the radius is 18.06; times π [3.14] is 56.7; that times the 10″ stroke is 567 cubic inches.)

645: designation of the engine introduced by EMD in 1965. The cylinder bore is 9¹⁄₁₆″ and the stroke is 10″, giving a cylinder displacement of 645 cubic inches.

710: designation of the engine EMD introduced in 1984. It has the same bore as the 645, but the stroke is lengthened to 11″. Cylinder displacement is 709.5″.

INDEX

UPDATE

As this book went to press in February 1989 there were several interesting developments in the diesel locomotive field.

Late in 1988 Burlington Northern commenced an ambitious Capital Rebuild Program, the first such program to involve Electro-Motive's La Grange plant. As initially announced, the program will upgrade GP30s, GP35s, and GP40s to Dash Two specifications. The rebuilt GP30s and GP35s will be rated at 2300 h.p. and called GP39s; the rebuilt GP40s will remain at 3000 h.p. but will be labeled GP40-2s. At press time, the roster of completed units was:

Numbers	Old model	New model	Rebuilder
2750-2759	GP30	GP39E	EMD
2800-2808	GP30	GP39M	M-K
2875-2880	GP35	GP39M	M-K
2925-2932	GP35	GP39E	EMD
3500-3508	GP40	GP40M	M-K

To avoid the need to take operating locomotives out of service, secondhand units — from Southern Pacific, CSX, and Union Pacific — are injected into the program as the work progresses. The potential for the program is at least 250 units if only BN's own units are rebuilt. The upgraded units should be as significant a feature of BN's roster as the Paducah rebuilds were on the Illinois Central-Illinois Central Gulf roster in the 1970s.

In addition, at press time BN took delivery of a trial order of Caterpillar-repowered GP20s.

Late in 1988 Canadian National's Point St. Charles Shops began rebuilding 15 A1A-trucked GMD1 light road-switchers. They are being retrucked to a B-B wheel arrangement, receiving new 12-645 engines, and being reoriented to operate short hood forward. The new class GR-412a indicates 1200 h.p.

In late 1988 GMDD began production of the F59PH, the cowl unit that will replace the F40PH in the 710-engine line. The 12-cylinder 710G engine is rated at 3000 h.p. in this model. A V-8 Model 149T engine at the rear of the unit drives a 500-kw, 575-volt AC HEP generator. The F59PH is 2 feet longer overall and in truck center distance than the F40PH. At press time, all production, an order for 22 units, has been for Government of Ontario Transit. Photo by Brian C. Nickle.

F59PH

SD60M

EMD's first Comfort Cab unit for a U. S. railroad is Union Pacific SD60M No. 6085, delivered January 19, 1989, as the first of a 25-unit order. If subsequent production of EMD's 60 line uses this cab, the face of American railroading will change considerably during the 1990s. Photo by George Cockle.

SD40F

In late 1988 Canadian Pacific began taking delivery of 25 SD40Fs, 9000-9024, built at London, Ontario. They have a cowl carbody with "Draper Taper" and the current cab configuration, but they have 645 engines, presumably for compatibility with CP's fleet of 560 SD40s. CP was dissatisfied with the units and sent them to EMD's La Grange plant for modification. Photo by Ian Platt. A footnote to the Caterpillar repowering section on pages 229-231: At press time CP was testing Caterpillar-repowered M-636 No. 4711.